The Cavalier Mode
from
Jonson to Cotton

The Cavalier Mode
from Jonson to Cotton

BY EARL MINER

PRINCETON UNIVERSITY PRESS

PRINCETON, NEW JERSEY

1971

L.C. Card: 74-155002

ISBN: 0-691-06209-9

Publication of this book has been
aided by the Whitney Darrow Publication Reserve Fund
of Princeton University Press

Printed in the United States of America by
Princeton University Press

For

THE STAFF OF THE
CLARK LIBRARY

In gratitude for their
unfailing kindness and assistance

PREFACE

THANKS TO its inherent value and the sensitivity of even casual readers of verse, Cavalier poetry has remained, in Waller's phrase, forever green in song. To anthologists, a garland of Cavalier poems has proved as necessary as are flowers to a wedding. The beauty of many of these poems has always, it seems, been recognized. Appreciation of the strength of Cavalier poetry has come more recently, both in superior new editions and in the excellent although still not numerous studies by critics and scholars. My hope is to add to these studies by an attempt to discriminate Cavalier poetry from other seventeenth-century alternatives, and to discriminate the major features within it.

Many useful approaches to the subject exist. As I said in the Preface to the study preceding this, *The Metaphysical Mode from Donne to Cowley* (Princeton, 1969), I myself value numerous levels of generalization. Our understanding can only be heightened by studying Ben Jonson in the general European context of his day, whether that context be termed Renaissance, Baroque, or something else. We have also long benefitted from the work of scholars like Ruth Wallerstein who preferred to work within a conception of the "seventeenth century." And as readers, teachers, or critics, all of us respond to the proposition that we must value individual poems. My present aim, however, lies between the grander and the closer views. As in my earlier study, so here I hope to describe conceptions of the self, of life, and the world held by poets in the late sixteenth and seventeenth centuries, conceptions which one group, the Cavaliers, tended to set forth in terms of certain styles, certain recurring subjects, certain recurring approaches, and certain cultural assumptions. More simply, it is a choice of human values and of artistic means. It will be plain that I have reason to draw comparisons between a Donne and a Jonson, because they shared a great deal as contemporaries and poets. And yet their values and their art are centered very differently within the larger perimeter. Moreover, this book is a sequel or a parallel to *The Metaphysical Mode from Donne to Cowley,* and I hope by such comparisons to draw the two accounts together without jostling the reader's elbow tiresomely. In both these books I have also looked ahead toward Milton and Dryden, again to draw resemblances and to discriminate within the yet larger perimeter of seventeenth-century English poetry as a whole. Also, I hope to conclude my survey of that century of poetry with a third volume in which Milton and Dryden will largely figure, and I hope that it, too, will not be without connection with my earlier accounts.

In the Postscript the reader will find a review of the ground covered in this book and some account of the rationale behind its ordering. This step seems advisable because Cavalier poetry is, after all, comparatively unfamiliar outside the usual anthology pieces, and because the reader may wish for retrospection after the numerous topics and numerous quotations necessary in treating a less familiar subject. But I owe it to the reader to speak briefly of the prospect before him.

The six chapters can best be grouped into three major divisions. The first consists of the first chapter, "The Social Mode," in which my aim is to describe the stance taken by the poets. Or to change the metaphor, I seek to delineate the results possible to a certain angle of vision upon oneself, one's subject, and one's readers. The next three chapters comprise the second division, an examination of the major values and principal concerns of Cavalier poets. The second chapter therefore introduces the central value, "The Good Life," considered both in terms of virtue (*vita bona*) and happy enjoyment (*vita beata*). The third chapter discusses the major threats to the good life. Its title, "The Ruins and Remedies of Time," suggests that the threat posed by time and the sorry times could also be met. And the ways in which the good life (in both versions) could be expressed as well as the means of dealing with various threats, especially those from the times, are examined in the fourth chapter, "Order and Disorder." The last two chapters make a closely related division of two, because the subject of the fifth, "Love," and that of the sixth, "Friendship," were constantly related and often distinguished on the same topical scale in seventeenth-century writings. In practice, however, Cavalier love poetry and poetry of friendship make up two large canons, and that factor, as well as the difficulty of many aspects of the subject, led me to treat the larger dual subject in two separate but related chapters. The division also makes it possible to use the last chapter somewhat as a gathering-in of issues and topics raised earlier.

In historical chronology, Ben Jonson begins this book and Charles Cotton ends it. Both are truly *sui generis* in the sense that each possesses his own vision of life and a capacity to transform the vision into poetry. If the book were designed with separate chapters on individual poets, it would honor that quality in their poetry far more, and for such failure I plead guilty before their *manes*. But chapters on individual author canons would involve tiresome repetitions of the same topics. One thing I have tried to do, at least after the opening pages, is to bring before the reader clusters of poems by individual poets at successive stages of my discussion. Sometimes Carew, or Waller, or Suckling, or Herrick, or yet another will be the representative man and object of poetic discussion, although poetic examples to illustrate the topic in hand might have been taken from a variety of poets. The reader will quickly discover that Jonson's role in this book is central, a fact which of course reflects my view of his importance. My affection for Charles Cotton will account for his relative prominence in the book. The other authors, some of whom are weightier than Izaak Walton's friend and all of whom are lighter than Jonson, will be found, I think, to repay close scrutiny. I shall ask indulgence on one score. I have shown less compunction than I would think desirable in a book on the Metaphysical poets to repeat quotations or to return to the same poem. And I shall think my chief aim a success if this book should lead to any revival of interest in the Cavalier poets.

The Clark Library   E. M.
Spring, 1970

ACKNOWLEDGMENTS

No BOOK I have written has closer ties than this to UCLA. Its ideas have grown from undergraduate and graduate teaching; its birth more or less coincided with a graduate seminar held at the Clark Library. I have long valued the unfailing kindness and assistance of the staff of the Clark Library, but I have never before written a whole book there nor have I ever found one grow to such an extent out of teaching. The other students in the seminar will not object to my mentioning Satendra Khanna for his work on Charles Cotton and David Latt for his work on friendship. We can claim to have taught each other, or to have helped each other with information we daily discovered at the Clark Library or the Huntington Library. There has been more than a little of the Cavalier rites in all this, even including drinking wine before my hearth in a (for California) heavy winter.

I also owe David Latt and Michael Seidel my thanks for checking my quotations and citations. For funds supplying such assistance and other help, I thank the Committee on Research at UCLA. I must also acknowledge the help received from the staff of the Huntington Library, including the warmth and courtesy of its director, James Thorpe.

Mrs. Jeanette Wallin has once again assisted me in the reading of proof. With the iteration of my thanks to her, I must include those to Frank Spellman for similar assistance in proofreading.

I wish to thank the Clarendon Press, Messrs. Routledge & Kegan Paul, Harvard University Press, W. W. Norton & Company, and the University of California Press for permission to quote from editions published by them (see Major Editions Used and Consulted).

Mrs. Eve Hanle has won my deep appreciation for her intelligence and insight in editing the manuscript.

ACKNOWLEDGMENTS

Although notice of the fact will be found elsewhere, I am particularly grateful to my friends Philip R. Wikelund and Thomas Clayton for advance texts of poems by Waller edited by the former, and poems by Suckling edited by the latter. Further information about the texts used, and the principles of modification, will be found in the section "Major Editions Used and Consulted." As this book entered the press, the Clayton-Beaurline edition of Suckling's poems, letters, and plays had not yet come to hand. It was therefore a courtesy very much appreciated that the Clarendon Press should send me revised page proof upon my receiving Professor Clayton's permission, and for this courtesy I thank the Delegates of that press as well as the editor. Professor Wikelund has gone to the length of collating, out of his intended order of work, some poems or passages in order to give me in advance his definitive text of Waller. I know that the reader will appreciate the fact that he has for the first time before him the only sound text of some of Waller's poetry; and he will understand what I mean when I say that such kindness is not so much an unusual scholarly courtesy as an act of special friendship that one can better appreciate than adequately acknowledge.

It is a pleasure to acknowledge yet again the assistance of two other good friends. Philip Levine has once more assisted me in matters classical, and James M. Osborn has shown his usual generosity in making material available.

I must also express my gratitude to the two readers of my manuscript for Princeton University Press. Much of what they said was most useful for its particularity; but I take it as the highest form of professional conduct and compliment that they should have held me up to the standards of my subject and of my own work. If they read this book, they will observe how useful this advice has been.

TABLE OF CONTENTS

CONTENTS

The Cavalier Mode
from
Jonson to Cotton

THE SOCIAL MODE

The mysteries of manners, armes, and arts.
—Jonson

THE SOCIAL mode is as much a radical feature of Cavalier poetry as is the private of the Metaphysicals or the public of
Milton, Dryden, and Pope. And by "social" I mean an aesthetic stance very much between those of Donne and of Dryden in their characteristic verse. The stance of the Cavaliers
is not one preventing them from writing of themselves and
those near them. In fact, they are in many respects more revealing about themselves than are their intenser Metaphysical
contemporaries. Jonson names (directly or indirectly) a son
and a daughter in his poetry; Donne, who had children to
spare, does not mention them. Herrick names his maid and
describes, often in great detail, features of country life. From
Jonson to Cotton, Cavalier lyrics abound with those dim
shapes, the Celias, Corinnas, and other ladies of classical
names, and often of real biographical existence. To have a
name, even a pastoral one, is to have the first rudiment of
social existence. Such names of the living will not be found
in the characteristic Metaphysical poems, nor will the relations of parents and children, friends, or subject and king. On
the other hand, such relations are regarded by the Cavaliers
in small societies, rather than in Dryden's public forum.

i. The Social Voice

The unnamed women pursued or argued with in Donne's
Songs and Sonnets are not poetically unreal because they lack

3

names. But their reality is of a kind different from that in the poems we are considering. To demonstrate the social mode and the nature of the reality of the characters, the *socii*, it will be useful to take a poet less well known (though very deserving of being known), Charles Cotton. By looking briefly at a few of his poems with reference to well-known poems by earlier Cavaliers, we should gain a sense of what the social mode is. Certainly, if we do not at the beginning catch the tone of voice of these poets, if we do not, as it were, know how to behave in their presence, and if we cannot allow the right kind of intimacy with the right kind of distance, then we shall find difficulty in understanding them. In Cotton's *Poems on Several Occasions* (1689), there occurs a succession of three poems representative of common Cavalier preoccupations. The first is a lyric epistle to a friend, "To my dear and most worthy Friend, Mr. Isaac Walton" (pp. 114-16). Cotton is chilled by winter in the north of England, but

> If the all-ruling Power please
>   We live to see another *May*,
> We'll recompense an Age of these
>   Foul days in one fine fishing day . . .
>
> And think our selves in such an hour
>   Happier than those, though not so high,
> Who, like Leviathans, devour
>   Of meaner men the smaller Fry.
>
> (21-24, 37-40)

Cotton sounds the true mid-century note. Outside there is cold and suffering, but warm days will come again, God willing, when we shall create a happiness better than that of the tyrannic great. And the happiness will allow us to re-create society, for we shall be together as anglers. The poem is not simply a Royalist attack on the Commonwealth or Protec-

4

torate. On the other hand, Cotton no more precludes such an attack than does Lovelace in "The Grasse-hopper."

> Thou best of *Men* and *Friends*! we will create
> A Genuine Summer in each others breast;
> And spite of this cold Time and frosen Fate
> Thaw us a warme seate to our rest. . . .

> Dropping *December* shall come weeping in,
> Bewayle th'usurping of his Raigne;
> But when in show'rs of old Greeke we beginne,
> Shall crie, he hath his Crowne againe![1]

The "best of *Men* and *Friends*" was Charles Cotton the elder, who helped warm the wintry days of the Interregnum by kindness to his friends, Lovelace says in the poem. But full union is not possible until the King, the bishops, and the old celebration of Christmas (banned by Parliamentary ordinance on 3 June 1647) come back again. The younger Charles Cotton seizes on May, which also had its rituals banned by those for whom the readiest name remains "Puritans." Cotton passes on to Walton a wisdom inherited by his father from Lovelace; and Walton, who had known Donne, Hooker, and other worthies, called Cotton his son and friend. The interlinking of friendly societies among the Cavaliers garlands the age, and we move from Jonson's Tribe or coteries at the Inns of Court, to Falkland's circle at Great Tew, to Waller and the Sidneys, back again to Jonson writing at the Sidney estate. Similar circles, or spirals, would take us forward to Cotton writing to Walton during the Restoration on a new edition of the *Lives*—the poem which in fact brings the last chapter of this book to its close. But such movements might also take

---

[1] Lines 21-24, 29-32. D. C. Allen has discussed the political implications of the poem in "An Explication of Lovelace's 'The Grasse-hopper,'" *Modern Language Quarterly*, XVIII (1957), 35-43.

us in numerous other directions, all of them governed by little, joined societies of friends.

The poem hoping for "one fine fishing day" is followed by one "To the Countess of Chesterfield, on the Birth of her first Son" (pp. 116-17). A mere country gentleman, Cotton is able to banter in the right tone with milady.

> A more wish'd for Heir by Heaven
> Ne'er to Family was given,
>   Nor a braver Boy to boot;
> Finer ne'er was born before him,
> One may know who got and bore him,
>   And now a-days 'tis hard to do't.   (13-18)

The superlatives of the first two lines of the stanza are cut to size by the rhythm and diction. "Nor a *braver* Boy *to boot*"— there is just the right, handsome, negligent, well-bred tone. Cotton concludes:

> You Copie well, for which the rather,
> Since you so well have hit the Father,
>   Madam, once more try your skill
> To bring of th'other Sex another
> As Fair, and Good, and like the Mother,
>   And double 'em after when you will.   (19-24)

Something of the man associated with *The Compleat Gamester* spices those lines, and something of the friend of gentle Izaak Walton sweetens them.

It is as if, in the right kind of world, life is a game, and a pleasant one. But the game is also that of life, when heirs may be brought forth "Fair, and Good," or even doubled "when you will." Less happy alternatives do not go unrecognized. A generation or two before, Jonson had written in similar terms about the Sidney family in one of his finest poems, "To Penshurst."

6

These, *Penshurst*, are thy praise, and yet not all.
   Thy lady's noble, fruitfull, chaste withall.
His children thy great lord may call his owne:
   A fortune, in this age, but rarely knowne.
They are, and have beene taught religion: Thence
   Their gentler spirits have suck'd innocence.
Each morne, and even, they are taught to pray,
   With the whole household, and may, every day,
Reade, in their vertuous parents parts,
   The mysteries of manners, armes, and arts.

                       (89-98)

The collocation of "armes, and arts" conventionally suggested
the epic ideal, and we recall that Sir Philip Sidney was a
soldier as well as a poet. The word "manners" bore a meaning
in the seventeenth century that it no longer has: habitual
moral conduct. The word therefore reveals an ideal of moral
integrity as well as social forms. "The mysteries" is a phrase
perfect in its inclusiveness of civilized, social, and religious
meanings that range from the "mysteries" of a London guild,
to secret rites of an initiate group, to the "mysteries" of the
Christian religion. In import, Jonson's great line occupies as
it were the realm between the crafts and the divine mysteries,
but with an implicit awareness of them not unlike Lovelace's
and Cotton's of politics in poems to friends. "The mysteries
of manners, armes, and arts" are precisely the civilized social
forms that much Cavalier poetry is about: provided we can
understand, as I hope this book shows, that a phrase like
"civilized social forms" allows for both lightness of touch and
moral weight.

   The third poem in this sequence of Cotton's is "To
Chloris. Stanzes Irreguliers" (pp. 118-21). Here we have one
of our classical nymphs and what many would think the true
tone of Cavalier wit.

> Lord! how you take upon you still!
>   How you crow and domineer!
> How! still expect to have your will,
>   And carry the Dominion clear,
> As you were still the same as once you were!
>
>   (1-5)

The assurance of tone derives from assurance of language and address. What is so right might easily have been spoilt into gaucherie or insipidity by the wrong emphasis on either verb in "How you crow and domineer!" Her fault in behaving in such ways emerges from the heavy accusation that she no longer bears the beauty of other days. But in the social views of the age, it was as unnatural for women to dominate as hens to crow. The poem depends, however, not on any single felicity or tone, but on a movement as problematical as that of a man's view of a woman's mind. So *he* will crow.

> I am now Master of the Gate,
> And therefore, *Chloris*, 'tis too late
> Or to insult, or to capitulate.   (20-22)

Here is a Cromwell at Drogheda, ready to put the enemy to the sword without a second thought.

Second thoughts provide, however, the motive force of the poem. At this point in his argument, the man turns (with results he fails to anticipate) to recall their amorous past.

> It was your Cheek, your Eye, your Lip,
> Which rais'd you first to the Dictator-ship.
>
>   (29-30)

Chloris had been the Cromwell, the Caesar, and he must confess his ignominious and "woefull Bondage." The fifth stanza brings him back to his claim.

8

But your six months are now expir'd,
   'Tis time I now should reign.   (31-32)

His dictatorship does not last long. If only she will be decent,

And love and honour me as I did you;
   That will an everlasting peace maintain.
                    (36-37)

There is something touching about his willingness to concede when he has his formerly cruel, and formerly beautiful, mistress in his power. The tone of the sixth stanza harmonizes the extremes of language and imagery into a full conviction of truth.

And Faith consult your Glass, and see
   If I ha'n't reason on my side;
Are those eyes still the same they use to be?
   Come, come, they're alter'd, 'twill not be deni'd.
                    (39-42)

We believe him. And we believe him when he goes on to say that she will refuse to: "For Womankind are all born proud, and never, never leave it." Everything so far has seemed to be true, suiting our knowledge of men and women and our emerging sense of the relations between these two people. And we are also convinced of human experience when he unexpectedly concludes with a deeper truth that he will continue to love her, and that his subjection in love is far from disagreeable. At the same time, his very hyperbole quietly reminds her of time and mutability.

And I must be a Subject still,
Nor is it much against my will,
   Though I pretend to wrestle and repine:
Your Beauties sweet are in their height,
   And I must still adore,

9

New years, new Graces still create,
Nay, maugre Time, Mischance and Fate,
You in your very ruines shall have more
Than all the Beauties that have grac'd the World before.
(50-58)

The movement of mind and feeling, the psychological experience, of Cotton's poem to Chloris recalls a number of Cavalier poems. Suckling's "Song" ("I prethee send me back my heart") is such another, and Waller has three poems resembling these: "To the mutable Fair," "To Phillis" ("Phillis 'twas love that injur'd you"), and the song, "Cloris farwell I now must goe." Perhaps some less devout admirer of the Cavaliers than I will say that he knew the Cavaliers all wrote alike. But the poetry most like Cotton's "To Chloris" is that of Cowley in *The Mistress*, where numerous pieces reveal the same divisions of mind, hesitations, shifts, sudden resolutions, and last decisions. Since *The Mistress* contains Cowley's most Metaphysical poetry, the charge requires a lawyer's inventiveness to be carried. The sinuous movement of this poem by Cotton and of many by Cowley indeed reminds us at once of such of Donne's poems as "The Indifferent." And yet a definition of "Metaphysical" that would admit this poem would be very ecumenical indeed. What fundamentally the poem possesses is a sense of the simple truth of the heart with another sense of the human complexity involved in trying to follow truth to its end. Cotton creates this double truth by admitting as much as possible of contrary motions. This is not irony or paradox. The end is not a dazzling Donnean wit or a poised Marvellian complexity. It is rather what has been very well termed a "catholic decorum."[2] The presence

[2] The term, which is exact for Cotton in one sense, and for seventeenth-century poetry at large in another, has been taken from the very perceptive dissertation by Satendra Khanna, "A Study of the Non-

of some kind of decorum sustains the social mode of Cavalier poetry; the catholicity, sometimes amounting to a surprising indifference to traditional ways of thought, gives Cotton his freshness.

If ancestors must be sought for such poetry, we can find them most readily in that many-sided poet, Jonson. In his "Elegie" ("Though Beautie be the Marke of praise"), as in Cotton's "To Chloris," there is continuous clarity of line with a movement, sometimes a poise, of attitudes that may bring a sense of mingled surprise and justice or of a mingled sense of justice with mystification. Donne dazzles us in one sense with his shifts of thought, although in fact his wit depends on a dialectic always stressing what we have shifted from, and what to. Jonson's poem begins with a stanza in which the concession is larger than the affirmation.

> Though Beautie be the Marke of praise,
>   And yours of whom I sing be such
>   As not the World can praise too much,
> Yet is't your vertue now I raise.   (1-4)

For six stanzas that virtue is celebrated, till finally the lady changes sex and changes from mortality to godhead: "And you are he: the Dietie / To whom all Lovers are design'd" (25-26). The lover who is speaking naturally becomes a worshipper.

> Who as an off'ring at your shrine,
>   Have sung this Hymne, and here intreat
>   One sparke of your Diviner heat
> To light upon a Love of mine.

Satirical Poetry of Charles Cotton" (University of California, Los Angeles, 1969).

Which if it kindle not, but scant
    Appeare, and that to shortest view,
    Yet give me leave t'adore in you
What I, in her, am griev'd to want.   (29-36)[3]

The "Love of mine" (32) designates a second woman, a fickle, contemptuous mistress; and this reading, like that "her" in the last line enables us to understand why it is the first woman's "vertue" that he praises, rather than her beauty. With the right tone he praises one woman he does not pretend to love in order to convey his suffering over loving another beauty without virtue. In such a delicate situation (and that it *is* delicate many a man has discovered to his cost), the extraordinary and therefore unexpected change of the good woman into a man and a god is a crucial metamorphosis enabling the poet to get away with worshipping a woman he does not love. And the worship of a god one does not adore surely aims at the relief of telling one's unhappy story, and maybe getting *some* kind of help in loving a beautiful woman who lacks both virtues of yielding and constancy. The tone is pure, the social relation sound. The poetic experience is by no means simple.

The complexities are essentially those of social relations interwoven with personal relations, and it is just this that distinguishes the aesthetic mode of the Cavaliers from that of the Metaphysicals. The sense of the desirable presence of others, the use of political allusiveness (and allusion *not* used as normally in the Metaphysicals, to stigmatize or to represent something altogether different: "She' is all States, all Princes, I'"), religious ceremony, the question of subordination of one sex to another—such elements are comprised in

[3] The first line of the passage reads "off-spring." I follow Whalley's emendation, "off'ring."

12

a poetry far removed from Marvell's garden with its flight from society:

> Society is all but rude,
> To this delicious Solitude.
>
> *(The Garden,* 15-16)

The "all but" is true Marvell in its reserve, but by comparison with Cotton (who is of course not the superior poet), Marvell seems to talk to himself, not needing in his more private and less social place to strike a tone of actual speech. Cotton has it.

> Lord! how you take upon you still!
> How you crow and domineer! . . .
>
> Madam, once more try your skill . . .

The very forms of address bespeak that kind of informality that only a certainty of forms can allow. Or, to retrace our tracks to the first poem introduced, the lyric epistle to Walton, we appreciate Cotton's seriousness, because we feel that the social tone, the metaphors, and the diction have suggestions of an actual social or political theme. Cavalier poetry would be very different without its social element, and in fact would often be difficult to distinguish from Metaphysical poetry at the one pole and Restoration poetry at the other. One can see as much in certain poets. When in mid-century those poets who were of Royalist persuasion were forced in upon themselves and turned by necessity away from the social scene they longed for, the poems they wrote sometimes enlarged the Cavalier stock but sometimes developed Metaphysical strains. The love poems of Cleveland and Cowley are as much, or more Metaphysical, than Cavalier; and in much of Marvell's best poetry the two lines of earlier seventeenth-century poetry merge.

13

The social quality was obviously not hostile, either, to that private, intimate world of the Metaphysicals or to the public, shared world of Milton and Dryden. Although the evidence is plain in Cowley, Cleveland, and Marvell, it will also be found, in varying degrees, in the works of other Cavaliers. Jonson has poems that his and Donne's editors dispute over and, as well, a goodly file of poems and masques opening on the public world. In other words, the social mode involves, in its general configuration, a mid-aesthetic distance, a position between the world of the poem and the world of the reader from which the poet can readily turn toward public poetry for certain needs, and toward private for others. Private poetry turns its heel upon the world, or affects to do so, in order that it may treat the transactions of the intimate heart. Public poetry avoids what is eccentric to the individual alone in order to celebrate what men share. So much for theory and the cruder features of extremes. In practice, numerous compromises and even thefts are possible between the two, and one of the happiest compromises is precisely the social mode.

Cavalier poets inherit the social mode from the Elizabethans. Jonson and his followers might fairly have entered claim to have been the true successors of a line of poets from medieval to modern times. There is a sense in which the private poetry of the Metaphysicals and the public of Milton and Dryden divagate equally from the traditional mainstream. It should go without saying that the best tuned instrument requires abilities in the player if the result is to move us. But that gift for song, which has been so much prized and sentimentalized over, was something Tudor poets possessed by virtue of a special adjustment of the individual to his world in which both elements could exert their claims, and this adjustment is that of the mid-aesthetic distance, of the social mode. If we take the Elizabethan sonneteers as a group, or the pastoralists, or the narrative poets, we observe that whether they treat the private experience of love or the public matter of good government, they do so by assuming, practicing, affirming social conventions. There is a fundamental sense in which Petrarchanism was the good form both of Tudor society and Tudor poetry. Whether a Courtly Maker, a University Wit, or an apprentice newly indentured to the Muses, an Elizabethan poet affirmed the bonds that held together a highly divided society. We see the same instinct in music, education, and religion. But it is also true, as every undergraduate shows in the exaggerated response of his ingenuousness, that the Tudor poets found it difficult to affirm social bonds and conventions without at the same time delivering themselves to the fetters of convention.[4] And the Cavalier poets who set forth from the late years of Elizabeth to remake poetry within the assumptions of the mid-aesthetic distance and the social mode faced just the same problem of affirming social ties without succumbing to literary convention. In this, the Cavaliers were as lucky to have the Metaphysicals as models of divagation as the Metaphysicals were lucky to have the Cavaliers as models of purity.

ii. The Creation of Person and Place

One purpose of this study is to show how the best Cavalier poetry works, how it enlivens conventions, whether social or literary, and how, to a reader grown accustomed to the accent of this province of English poetry, it has something enduringly important to say. As Charles Cotton had served my purpose earlier to convey something of the quality of the musical key, as it were, of the social mode, so now Edmund Waller may be taken to show something of the range of Cavalier

---

[4] This, and other, aspects of Tudor literature are well set forth by Hallett Smith, *Elizabethan Poetry: A Study in Conventions, Meaning, and Expression* (Cambridge, Mass., 1952), a book I, like others, particularly value.

motifs after Jonson. Waller's name is no longer the coffee-house word it was to critical Dick Minims in the eighteenth century, and his poetry has not been analyzed out of existence. To such negative reasons for choosing him, I may add the historical one that he did truly seem to another generation of writers to have played a significant role in the development of poetry in their times. I also like him. Waller's "Of the last Verses in the Book" provided him with an opportunity to be personal in the manner of a Roman poet (or Chaucer) bidding his little book farewell. Instead he seems about to write a public poem.

> When we for Age could neither read nor write,
> The Subject made us able to indite . . .   (1-2)

Here and in the next ten lines we encounter the "we" that so distinctly marks public poetry. The last verse paragraph, however, triumphantly shifts to a middle, social mode; significantly, it provides the best poetry. The preceding two lines (11-12) may be given in order to mark the shift in quality and mode by a contrast.

> Clouds of Affection from our younger Eyes
> Conceal that emptiness, which Age descries.
>
> The Soul's dark Cottage, batter'd and decay'd,
> Let's in new Light thrô chinks that time has made:
> Stronger by weakness, wiser Men become
> As they draw near to their Eternal home.
> Leaving the Old, both Worlds at once they view,
> That stand upon the Threshold of the New.   (11-18)

What in Carew's poetry is so worthy of being termed Meta-physical in nature as those six lines? In spite of that, in spite of Waller's purity of tone, and in spite of his concern with "life," it was Carew that F. R. Leavis chose to dignify as one in "the line of wit." Very strange Leavis's choice may seem, although we must observe that Waller does lack a passionate private address, or rather that he speaks at most with a "wilde civility." The truth of the poem concerns not so much of *me* precisely, nor yet of *us*, but of *"Men."* Waller mediates among these implicit rivals with the social mode.

Another of his poems, this one certainly dating from before the Restoration, "At Pens-hurst" ("While in the Park I sing, the listning Deer") is in some ways yet more instructive, because its claims are so unabashedly Cavalier that we may forget the Metaphysicals, Milton and Dryden, or other alternatives and enjoy Waller.[5] But we are led to recall with this poem, and with another to which Waller gave the same title, Jonson's splendid poem. Unlike the other poem, however, this does not echo *To Penshurst*, and Waller truly adapts the motifs of topographical poetry with fetching subtlety. Sacharissa's wooer is out of doors, in the park at Penshurst, where nature responds to his song of lover's woe. She remains cold.

> To thee a wild and cruel soul is given,
> More deaf than trees, and prouder than the heaven![6]

Sacharissa resembles neither parent. Her descent is inhuman: "to no humane stock / We owe this fierce unkindness; but the rock . . ." (19-20). Such complaint may seem merely a heightened version of Petrarchan convention. What the convention (if this is that convention) requires is either a retraction, or some kind of recovery revealing that the lover had momentarily lapsed into madness. But not Waller (at least not until four unfortunate lines at the end). He proceeds almost directly on the hypothesis that the man has full right in

[5] See the Appendix for the text of the poem.

[6] Lines 7-8. The ordonnance and complexity of the couplet will be appreciated by a seventeenth-century view of nature and classical deities expressed in the opposition—wild (deaf) trees, cruel (prouder) heaven: you.

his cause. As a poet, he says, "I might like *Orpheus* with my numerous moan / Melt to compassion" that obdurate stone, Sacharissa. But "my traitrous song / With thee conspires to do the Singer wrong." His poetry fails to assuage his suffering, fails to make progress with her, and indeed fails in all respects except for providing poetry for him and us.

At this point Apollo enters,

> Highly concerned, that the Muse should bring
> Damage to one whom he had taught to sing.
>
> (35-36)

The tone seems perfect: the lover's problem has become Apollo's. And with the entry of Apollo's wrinkled brow, the poet (Sacharissa's lover) and we stand off in an interest mingling concern with amusement. The best the god of the caduceus can propose is departure from Penshurst for the mind-relieving "wonders" of the seaside. Life by the seaside held no attraction at all to the seventeenth century, and the sea itself was variously thought a desert or a treacherous medium, and not without good reason.

> Ah cruel Nymph from whom her humble swain
> Flies for relief unto the raging main;
> And from the windes and tempests do's expect
> A milder fate than from her cold neglect . . .
>
> (41-44)

The four lines following do in fact provide the conventional palinode, but, as their relative brevity shows, the effort proves perfunctory. They provide a possible basis for another poem and only mar the present one.

The first forty lines give us another, much superior poem. Sacharissa is out of harmony with the wooer, with nature, and with her family estate. She is "wild and cruel" like that sea which, ironically, is supposed to promise some relief to the wooer. Like the man, unlike the woman, the park is harmonious and sympathetic. Only that daughter of the rock (which might better be by the side of the wild sea), Sacharissa, is inhuman, unnatural, unmoved by art or even by the very god of art. Such extraordinary accusations are preferred without euphemism. But they enter in the right tone, one entailing self-respect and a conviction on the lover's part of the justice of his cause. A poem of this kind must avoid attenuation, and this Waller does do; but it must also avoid as well the contrary danger of excess, and here Waller ran his greatest risks. The danger was run in the accusations; it was avoided by the rightness of phrasing in rhythm and language. But how, the devotee of seventeenth-century poetry will ask, how is Waller's *mind* engaged?

"At Pens-hurst" obviously draws on the resources of the topographical, or "loco-descriptive," poem. Waller's originality lies in his fundamental rearrangement of elements. The encomium on the Sidney family is founded on that exception, Sacharissa, whose aberration provides the poem with its reason for being. The contrast between her and the place, like the sympathy between her lover and her place, renders the topographical poem into amorous complaint as much as the reverse. The reality of the place serves to prove the reality of the complaint: she should return his love. We do not find it difficult to observe such polarities as the wild/civil, in-nature/out-of-nature. We are probably less aware of the poem's basis in two topographical motifs that become explicitly thematic in lines 41-44. Estate poems and sea poems are two of the several kinds of topographical motifs.[7] Waller combines the two in a causal theme: the mistress differs so much from the

---

[7] See Robert Arnold Aubin, *Topographical Poetry in XVIII-Century England* (New York, 1936) and, with care, G. R. Hibbard, "The Country House Poem of the Seventeenth Century," *Journal of the Warburg and Courtauld Institutes*, XIX (1958), 159-74.

park that the lover must go to the wild sea for relief from her yet greater cruelty.

Attention having been given to the interests of Waller's poem in itself, we may reintroduce comparison with other poets. "At Pens-hurst" skillfully employs the motif of the eavesdropping god begun in love poetry (so far as my knowledge extends) by Ovid's first elegy (*Amores*, I. i), a technique often involving shifts in audience as well as in speaker, as Donne's poem "The Indifferent" well shows.[8] The three poems share a lively wit, an agility in maneuvering speakers and audiences that almost conceals the sleight of hand. With Waller, we can observe that the first response gained by the speaker comes from nature, which therefore proves to be his first audience (1-6). Then he turns, by contrast, to Sacharissa, addressing her (7-32) and obtaining even less positive response. With the entrance of "just *Apollo*, President of Verse," the lover as well as Sacharissa becomes the audience of the god, who tells him to go to the shore (33-40). In the last eight lines, the despairing lover addresses Sacharissa as (apparently) he is about to leave for the seaside. The shifts in the identity of the speaker and of the audience give the poem the liveliness it requires.

In its technical virtuosity, "The Indifferent" is one of Donne's most dazzling poems, and his dramatic effect is much stronger than Waller's in this poem. But unless I am much mistaken, comparison of these two poems alone shows Waller's to possess more of what we recognize to be human truth. There really is a significant bond in his social attitude. The conventional language is one symptom, and not the happiest. The playing off, throughout the poem, of mid-stopped and

---

[8] I have tried to discuss this aspect of Donne's poem in *The Metaphysical Mode from Donne to Cowley* (Princeton, 1969), pp. 15-18. The resemblance is a significant one, but I think Waller's poem depends for its effect on a stance more distant than Donne's "dramatic" procedure.

end-stopped lines is a better symptom, because it echoes in little a balance or restraint with a freedom that are essential together for social intercourse. Numerous details reveal how sure Waller is of his world: the atmosphere of a country seat whose very animals and plants make up a sympathetic audience (1-6); the mention of the family name (10) and of Sir Philip Sidney (11-14); and the easy introduction of "just *Apollo*, President of Verse" (with his due social title and a flicker of humor). What is of equal significance is that the world of which Waller is so sure is also a world that knows him and responds to him. Such assurance enables him to speak things about Lady Dorothy Sidney that neither Dryden nor especially Donne would dare say of their high-placed ladies.[9]

"At Pens-hurst" does not possess the same perfection, the same inevitability, as Waller's most familiar poem, the "Song," "Go lovely Rose." "At Pens-hurst" lacks such purity because it hazards more risks, which it handsomely overcomes—all but the very last. It seems closer to the Metaphysical line, some might say, although the differences are crucial. In Donne's poems, passion is defined but not circumscribed

[9] The biographical element in Cavalier poetry is in some ways even more difficult to allow for than in, say, Donne or Dryden. Sacharissa is a version of Lady Dorothy Sidney, as everyone knows, but no one knows how accurate a version it is (there is an account in Julia Cartwright, *Sacharissa* [London, 1893]), nor indeed how accurate Waller chose to be in describing his own feelings. There is an implicit caveat in the example of Robert Bell, editor of the *Poetical Works of Edmund Waller* (London, 1854). Bell had the temerity to order the Sacharissa poems into presumed chronological order and the assurance to say of Waller that it cannot "be assumed from his verses that his feelings were very deeply engaged" (pp. 19-20). Moreover, of the treatment of Sacharissa in this poem he says, in yet another contradiction, "the reproaches heaped upon her . . . are not creditable to the generosity of the writer" (p. 87). The "reproaches" are precisely the source of poetic energy, though I cannot pretend to untangle the biographical knots Bell ties himself into.

or controlled by awareness of it; and his speaker, usually a young man, feels his own amorous pulse and regards his woman for symptoms of her response. Any awareness of third parties entails their inferiority and his confirmed self-esteem. Waller's appeal to what is due himself entails limits, because it turns on an ethical question: ought Sacharissa behave so? Such qualification is very Cavalier, as is also the psychological qualification that Suckling so preeminently argues. Questions of ethical justice or psychological veracity necessarily appeal to an audience not eccentric in its norms, but rather one possessed of a *jus gentium*. This, the largest audience of the poem, includes us as part of that great world that is implicit throughout the poem and very nearly explicit in the contrast drawn between Sacharissa and her parents, or between her and Sir Philip Sidney who

> could so far exalt the name
> Of Love, and warm our Nation with his flame.
>
> (11-12)

Waller of course does not suggest that the "Nation" includes the total population, but rather those who value Sidney as much as he does—men and women of quality in a double sense. Such persons of quality make up the social milieu of Cavalier poetry, but since we can also find value in Sidney and answer ethical questions when they are posed to us, we too are part of the social world of the Cavaliers. That is, we have terms on which we can prize it. By the same token, we enter that world on Waller's terms, and these are terms of fiction or myth, since he represents himself singing like another Orpheus who affects animals and trees (1-4) and by introducing Apollo as a major audience and speaker. Alongside us in the listening audience there stand deer and beeches, and to us comes "*Apollo*, President of Verse." What we share with animals, trees, and the god unites us with the wooer and divides us from Sacharissa: a value of song and the passionate feelings of the singer.

Waller set a standard for lyric poetry to the end of the century, when his verses assisted Millamant and Mirabel in understanding themselves and each other. Waller of course lived into the Restoration and made up with others both the first and second of those two mobs of gentlemen that, Pope said, wrote with ease. Historically speaking, Cavalier poetry alone endures throughout the century. But one must admit that its canons become adjusted and its assumptions gradually yield to others. The songs or comparable lyrics by Restoration poets may sometimes be very Cavalier, as much of Waller or Cotton shows. But the alternative lyric form distances the situation by a greater fiction, often allowing for its existence only in the special context of a play, and often by a use of pastoral that does not invite us to identify the speaker of the poem with the poet. Only perversity would deny us an association of Waller and his speaker; and only perversity would associate Dryden with any of the speakers in his love songs. On the other hand, only perversity would deny us the association between Dryden and his speaker in such complimentary address to a lady as we find in *To the Dutchess of Ormond.* Dryden goes to considerable length to establish the nature of his relation to the Duchess, and what emerges must be accounted a genuine tie, and yet a tie that does not possess the equality of wooer and wooed, but a very different equality of the woman celebrated and the *poet* who immortalizes her. Sacharissa, like Sidney's Stella, may in the end lie beyond the wooer's reach, but the psychological and ethical concerns of Waller or Astrophil make them the norm in the experience of love, restoring the equality only seemingly lost in the Petrarchanism they share.

If Cavalier love poetry distinguishes itself by its characteristic development of Tudor attitudes in a new, more relaxed version of the social mode, it is the same social adjustment of poetry that enables us to distinguish Cavalier public poetry from the more fully public version of Milton, Dryden, and the Augustans. Later we shall have sufficient occasion to observe different strains of Cavalier poetry concerned with the times. In one, especially typical of Jonson, the ethical strain expresses itself in a degree of commitment so strong that we are taken to the utmost bounds and to a confession of personal ideals. Another ethical strain, especially common during the Interregnum, takes disengagement or withdrawal to the bounds of our bafflement, and sometimes beyond. A third ethical strain involves a degree of participation in a situation where other men and women may appear in groups. The third of course presents the closest approach to the Restoration mode and requires the closest discrimination in order that the resemblances, the differences, and the special achievements of the Cavaliers can be seen, and felt.

Waller's finest poem in the third ethical strain is *On St. James's Park as lately improved by his Majesty*. Charles II is the improver, but the situation of the poem resembles those in such earlier ones (with such awkward titles) on Charles I as: "Of the Danger His Majesty (being Prince) escaped in the rode at St. Anderes"; "Of His Majesties receiving the News of the Duke of Buckingham's Death"; "To the King On His navy"; or *Upon His Majesties repairing of Pauls*. The last of the poems was much the best known and often referred to in the century. Like the rest, it seems, in its cumbrous title, to wish to draw close to the subject but to wish the subject to stand at a distance. This element we find in *On St. James's Park*. We do not find it in Dryden, whose distance may vary but is always certain. The fact that Dryden joined others in praising Waller's poem on the efforts of Charles I to remodel St. Paul's[10] suggests that he responded to this strain in Waller,

10 Dryden, *Annus Mirabilis*, st. 275.

and that on occasion at least he felt that there was more to Waller than the refiner of the language, more than a poet of natural phrasing, and easy prosody.[11] Much in *On St. James's Park* could only have appealed to Dryden: the Royalism, the depiction of harmonious activity, the celebration of art and civilization. But the quality that would have appealed most to him is the quality affording the greatest interest today, a species of artistic unity that grows from admission of a maximum in detail and from a harmony combining form with dynamic movement.[12]

Waller begins by speaking of a paradise lost (1-2) and then of another regained. We have been taught recently to understand the workings of such *topoi* and their Royalistic implications in Dryden.[13] It is no small art that enables Waller to introduce the traditional figures with such naturalness. A topic seems to come with ease, and so does the next in an associative process. And yet, one remarkable feature of the poem resides in Waller's unwillingness to introduce any topic or image of importance only once. The regaining of paradise seems to depart with as good a grace as it enters; but when it later enters with the same grace, we are conscious that it represents but one note in a complex, lovely harmony. If paradise is to be regained, what we require is, in Milton's phrase, "one greater man," and that man is Charles, as Waller's concluding

[11] Dryden is so ready with praise that many have failed to observe the strict delimitations of his encomia. In the Preface to *Fables*, he praises the line of Fairfax, Denham, and Waller for what may be called clarity and wit; but his line of genius runs from Chaucer to Spenser to Milton and, no doubt, to Dryden. See *Of Dramatick Poesy and Other Critical Essays*, ed. George Watson, 2 vols. (London, 1962), II, 270-71. Our usual opinions have been strikingly reviewed by Paul J. Korshin, "The Evolution of Neoclassical Poetics: Cleveland, Denham, and Waller as Poetic Theorists," *Eighteenth-Century Studies*, II (1968), 102-37.

[12] For the text, see the Appendix.

[13] See Alan Roper, *Dryden's Poetic Kingdoms* (London, 1965), "The Kingdom of Adam," pp. 104-35.

lines (127-36) make clear. By the close of the poem, Waller
rises to magnificence. The star that shone in mid-day at the
birth of Charles II Waller applies (and not in the first poem
to do so nor the last) to the Star of Bethlehem, since being
"Born the divided world to reconcile," Charles plays on earth
a role like that of Christ on earth and in heaven. The last line,
which promises that the King will reform nations as he has
made "this fair Park from what it was before," clearly sub-
stantiates, validates the opening line of the poem.

The lengthier middle section of the poem (5-126) offers a
number of versions of what the Park's essential significance
may be. The versions seem discrete enough and sufficiently
amenable to description. For example, one version presents
the Park as a social gathering place.

> Me-thinks I see the love that shall be made,
> The Lovers walking in that amorous shade,
> The Gallants dancing by the Rivers side;
> They bath in Summer, and in Winter slide.
>
> (21-24)

In a few lines we encounter another version of the Park as a
natural scene.

> Whilst over head a flock of new sprung fowl
> Hangs in the ayr, and does the Sun controle:
> Darkning the sky they hover or'e, and shrowd
> The wanton Sailors with a feather'd cloud.   (27-30)

These two passages of four lines reveal how two versions of
the Park (as meeting place and as natural scene) relate. Each
is an aspect of the other. Lovers will meet: to walk in the
("amorous") shade.[14] Nature provides the gallants with a

14 That epithet, "amorous," for a "shade" reminds one of Marvell's
Garden. Whatever the dates of composition, Waller's poem was pub-

river to bathe in during the summer or to slide upon in the winter. Similarly, the newly fledged birds (which are surely waterfowl from the context) also assemble, shadowing "The wanton Sailors" visiting the Park. The sailors in the natural scene recall the Park as a meeting place, and the birds assembling in the air provide a yet subtler version. Moreover, as the sailors bring back the so-to-speak watery associations of the earlier scenes, the birds "shrowd" the sailors also as if the fowl were sails. In such fashion, Waller merges detail that we apprehend distinctly in our immediate encounters and yet that accumulates in our memories of what may be termed, with some appropriateness for this poem, the spatial and temporal dimensions of our experience of reading.

The Park also possesses a third version, the mythic. Speaking of the trees that have been newly planted, Waller says,

> The voice of *Orpheus* or *Amphions* hand
> In better order could not make them stand.
>
> (15-16)

Orpheus and Amphion of course produced their effects on trees or the walls of Thebes through their music, and a later couplet, falling between the two quatrains on the Park as meeting place and as natural scene, picks up the musical implications of the myth and joins the two quatrains by writing of "water-music":

> Me-thinks I hear the Musick in the boats,
> And the loud Eccho which returns the notes.
>
> (25-26)

Waller does not *mention* the music of Orpheus and Amphion (it is rather the "voice" of the one and the "hand" of the

---

lished two decades before Marvell's, and it is well known that Waller is one of the poets that Marvell echoes with some frequency.

other), any more than he draws an explicit connection between gallants in the river, boats, boat music, and "wanton Sailors." But one comes to marvel at the ease, that negligent inevitability of Waller's art. The Orpheus and Amphion legends are of course used negatively, in the manner of Milton: the classical artist-creators could not surpass this scene.

Such discrimination introduces the middle section of the poem.

> Instead of Rivers rowling by the side
> Of *Edens* garden, here flowes in the tyde.
>
> (5-6)

Equal to the best of antiquity, the Park is not the same as the first paradise, Eden. But the Park represents of course the paradise lost of the preceding four lines and that regained at the end.

One experiences no difficulty in identifying the superficial details of the "mythic" version of the Park. One quickly finds "Cupids" (37), "*Noahs* Ark" (43), "*Peters* sheet" (44), a "muse" (67), "sacred Groves" (74), "*Romes* Capitol" (88), Augustus (123), and Hercules (124). Some of this detail may well seem "the goodly exil'd traine / Of gods and goddesses" that Carew credited Donne with banishing from "nobler Poems," and all of it fits into something that might be termed lore of the Stuart myth. Certainly Waller never shirked a classical reference simply because it was familiar. But when the details fit, and when they make larger wholes, it does not matter whether the familiar detail is Hercules or a skylark. Charles II is meditating

> What Ruling Arts gave great *Augustus* fame,
> And how *Alcides* purchas'd such a name.   (123-24)

Augustus is a good model, obviously, for one who would practice the "arts" of ruling in peace. But Hercules may seem less to the point. By long tradition, however, Hercules was a type of the king, and especially "the type of a good king, who ought to subdue all monsters, cruelty, disorder, and oppression in his kingdom, who should support the heaven of the Church."[15] Augustus is a type of the wise, politic ruler, and Alcides of the active, so that Charles as he meditates seeks to reconcile in an ideal or a mean the highest standards of kingship. Details that seem conventional may be so indeed, but then I doubt that Donne was the first to observe that it is exciting to make love or Dryden the first that a nation thrives better in concert than in anarchy. Waller, that is, earns his right to use the common treasure of humanism.

Waller's "Park" has echoes and significant detail. The poem also employs a movement at once forceful and subtle. A comparatively simple example of such movement can be given by reference to the progression of the shadow images that we have already seen. First, we shared Waller's vision of "The Lovers walking in that amorous shade" (22). Next, we observed the eclipse by birds:

> Whilst over head a flock of new sprung fowl
> Hangs in the ayr, and does the Sun controle:
> Darkning the sky . . .   (27-29)

That is fairly breathtaking. It also gives every reader a slight sense of threat, a threat that dominates in a later and seasonally contrary phrase, "Winters dark prison" (53). Such a threat must be overcome, and if it is too early in this book to emphasize the political significance of winter to the Cavaliers, we can understand Waller well enough by his shades.

> Near this my muse, what most delights her, sees,
> A living Gallery of aged Trees,

---

[15] Alexander Ross, *Mystagogus Poeticus* (London, 1647 *et seq.*), *s. v.* "Hercules." For some further lore, see *The Works of John Dryden* (Berkeley and Los Angeles, 1969), III, 305.

Bold sons of earth that thrust their arms so high
As if once more they would invade the sky;
In such green Palaces the first Kings reign'd,
Slept in their shades, and Angels entertain'd.   (67-72)

Such lines require no praise, but they may be given some explanation. The easiest explanation will be found in James Howell's Δενδρολογία. *Dodona's Grove, or, the Vocall Forest* (London, 1640), an allegory of the parlous state of England in terms of talking trees. As Howell says "To the Knowing Reader,"

Then be not rash in censure, if I strive
An ancient way of fancy to revive,
While *Druyd* like conversing thus with *Trees*,
Under their bloomy shade, I Historize:
Trees were ordaind for shadow, and I finde,
Their leafs were the first vestment of Mankinde.

(sig. A 2ᵛ)

In 1650, Howell had his second part of Δενδρολογία brought out. The King had been executed more than a year before, and this staunch Royalist has a sad allegory to relate. Indeed, he ends with a prayer that explains philosophically, or emblematically, why Charles II had to improve St. James's Park.

I Will conclude with my incessant orisons to Heaven that the All-powerfull Majesty of the Univers, without whose providence a *leaf* cannot fall; Hee who can turn *Forests* to *frank-Chases*, *Chases* into *Parks*, *Parks* into *Warrens*, and all four into a *Common*, and in lieu of *Hart*, *Hind* and *Hare*, with other harmless Cretures, can fill them with *Tygers*, *Beares* and *Wolfs*; May that All-disposing Deity who can turn *Empires* into *Optimaces*, *Optimaces* into *Democraticall* States, *Democracies* into *Oligarchies* at his pleasure; May that high Majesty, I pray, vouchsafe to put a period to these black Distractions in *Druina* [i.e., England] by an indissoluble bond of a *setled* Peace, and not turn the light of his countenance quite away from her; lest according to the words of the holy Prophet, *The Trees of the Forest becom so few that a Child may write them; lest the places where the green reed and rush doth grow, becom dens for Dragons,* and, which is justly to be feared, lest the whole *Forest* becom a *Beggars Bush.* (pp. 286-87)

And there follows a picture of the robust author, in fine Cavalier garb and brown study, leaning on a tree, "*Robur Britannicum*" the British Oak. Or to put this lore of trees in brief, we may recall the motto of both parts of Howell's arborial allegory: "Let the tree be honored whose shade protects us." Or, in view of the "darke conceit" of the work and perhaps implied by "shade" (*umbra*), "Let the Oak be honored, whose royal shade protects us."[16]

Waller's poem of course strives for no allegorical shadows, but it does concern itself with history, natural or human, like Howell—or indeed like Martial. It is widely thought that the "loco-descriptive" poems by Jonson, Carew, and Waller have behind them one of Martial's longer epigrams, III. lviii. A comparison of Faustinus's villa in that poem with St. James's Park entails, really, only one point of resemblance: they are places of pleasure. No doubt there is an important sense in which St. James's Park is the country in the city (*rus in urbe,* Martial, xii. lvii. 21), but the poem by Martial that Waller most draws upon is IX. lxi (or lxii in older editions such as the Delphin [my copy, Paris, 1680]). Martial describes a Cordovan house at which Julius Caesar had planted a plane tree; Waller a park improved by the British Caesar. Both poems

---

[16] "Arbor honoretur cujus nos umbra tuetur," a proper hexameter. For similar poetic and political use of trees, see Herrick, "Farwel Frost, or welcome the Spring," ll. 5-22, also the Latin verses, ll. 1-3, on the frontispiece of *Hesperides.*

31

treat the crucial three images, trees, shade, and the stars or heavens (*sidera celsa*; Waller, 69-70, 128). Both poems involve myth and religion with a historic moment. Numerous differences exist between them, and no doubt the political shade (*umbra*) of Howell is as much to the point as the festive of Martial, since Waller has an "oraculous shade" as well as an "amorous" one. In just such interplays between present urgencies and classical norms we find much of what is most characteristic of some of the most resonant poetry of the century.

Waller's lines respond, then, to seventeenth-century experience by means of the emblems and recollections by which the century chose to understand itself, and they do so in lines of unusual force. Every reader must feel that those arborial "green palaces" belong to the seventeenth-century poetry in all styles. In much the same vein, and within ten lines, we hear of "this oraculous shade" (80) where the world's fortune is decided. The phrase of course recalls the earlier "amorous shade," and the development of, as well as the qualification by, the earlier phrase in the later is very much in the witty cast of the century. The shadow imagery is not left in wit, however, but is taken to a transformation like that favored by the Cavaliers (see ch. iv, below). Both the passage earlier on the birds and that on the green palace of the trees had directed our attention toward the sky, and both had conveyed some sense of threat or violence. The danger is resolved, and the shadow imagery is transformed, by that star at Charles's birth:

> what the world may from that Star expect
> Which at his birth appear'd to let us see
> Day for his sake could with the Night agree.
>
> (128-30)

The shade is transformed but not dispersed; it is reconciled with the light.

Such reconciliation provides us with another version of the significance of this park royal. The Park resolves numerous contraries (including some implied earlier) by the principle of *discordia concors* (or *concordia discors,* the two phrases being used to mean the same thing by classical authors).[17] Waller's approach, at its simplest, involves stress on the evident reality: a park has a unity in its bounds and yet may be stocked with a variety of places, people, trees, and other inhabitants. But he takes pains to develop contraries, as for example of the gallants frolicking by the river: "They bath in Summer, and in Winter slide" (24). But even a talkative critic should let his poet speak for himself and should let his poet's images play their roles.

> Yonder the harvest of cold months laid up,
> Gives a fresh coolness to the Royal Cup,
> There Ice like Crystal, firm and never lost,
> Tempers hot *July* with *Decembers* frost,
> Winters dark prison, whence he cannot flie,
> Though the warm Spring, his enemy draws nigh:
> Strange! that extremes should thus preserve the snow,
> High on the Alps, or in deep Caves below.　(49-55)

> What Nation shall have Peace, where War be made,
> Determin'd is in this oraculous shade;
> The world from *India* to the frozen North,
> Concern'd in what this solitude brings forth.　(79-82)

[17] See the outstanding analysis of Denham's *Cooper's Hill* by Earl R. Wasserman in *The Subtler Language* (Baltimore, 1959), pp. 45-88. In his critical discussion of the poem, Brendan O Hehir, *Expans'd Hieroglyphicks* (Berkeley and Los Angeles, 1969), treats of the "harmony" in terms of diadic and triadic relations among "extremes" and similars.

From shades near Whitehall, the seat of government and the *imperium*, the King's vision directs us to Westminster Abbey and the *sacerdotum*.

> From hence he does that Antique Pile behold,
> Where Royal heads receive the sacred gold;
> It gives them Crowns, and does their ashes keep;
> There made like gods, like mortals there they sleep
> Making the circle of their reign compleat,
> Those suns of Empire, where they rise they set.[18]

Next, a glance at the Parliament, and a return to the Abbey.

> Hard by that House where all our ills were shapt
> Th' Auspicious Temple stood, and yet escap'd.
> So snow on *Aetna* does unmelted lie,
> Whence rowling flames and scatter'd cinders flie.[19]

And finally, the reconciliation of all in the King (and in Christ the King, who reconciled heaven and earth with His dual nature). Charles considers

---

[18] Perhaps I may be permitted, in a note, to comment on two things: the emblematic character of the passage, with gold and the sun for kings; and the continuity of such emblems as those of the circle in the seventeenth century. It is sometimes supposed that the circle was broken and emblems were dead as Pan in 1660: this poem was probably written and certainly published in 1661. See also *Absalom and Achitophel* (1681), l. 838; *To My Honour'd Kinsman* (1700), l. 65; and the Dryden *Concordance*, s. v., "circle," "circles," "sphere," "spheres," etc. Important changes took place in poetry and life in the seventeenth century, and there were numerous alternative views; but certain kinds of fanciful ideas about early *vs.* late provide more comfort to superstition than to truth.

[19] Lines 99-102. The Aetna simile derives ultimately from Claudian, *The Rape of Proserpine*, i. 163-70. But it was a frequent formula for *discordia concors*: see, e.g., Cowley, "To Mr. HOBS," st. vi, and Cowley's note on it.

what the world may from that Star expect
Which at his birth appear'd to let us see
Day for his sake could with the Night agree.
A Prince on whom such different lights did smile,
Born the divided world to reconcile.   (128-32)

Having let Waller more or less speak for himself, I shall add a few small amplifications. In line 43, for example, Waller speaks of "The choicest things that furnisht *Noahs* Ark." Among the significances attached to the ark is that of a unifying of great variety.[20] Much the same significance of variety reconciled inheres in the poem's first depiction of Charles, although that significance does not perhaps emerge clearly until we have seen the other passages of the poem.

Here a well-polisht Mall gives us the joy
To see our Prince his matchless force imploy:
His manly posture and his graceful mine
Vigor and youth in all his motion seen,
His shape so lovely, and his limbs so strong
Confirm our hopes we shall obey him long:
No sooner has he toucht the flying ball,
But 'tis already more than half the mall;
And such a fury from his aim has got
As from a smoking Culverin 'twere shot.   (57-66)

"His shape so lovely, and his limbs so strong." The uniting of beauty and strength (often figured in Venus and Mars) is a key reconciliation in the lore of *discordia concors*, for from that reconciliation was born harmony. The vision of a prince relaxed (the art of relaxation being one that Charles II practiced to perfection) yields to a simile for war. Charles's re-

[20] See, from 1656, Cowley, "ODE OF WIT," st. viii; from 1667, Milton, *Paradise Lost*, xi. 732-37, 892-901; and from 1685, Dryden, Killigrew Ode, ll. 123-26.

tirement to the peace of the Park does not exclude the outside world; there is wide allowance for what has been only apparently left behind (see ll. 76 ff.). It is no wonder that the next twenty lines concern both civil and foreign war. The passage introducing the King plays a crucial thematic role, defining as it does in most important terms the version of the Park as a meeting place, introducing the chief character of the poem, and showing how far the Park is defined by the role of the King.

The King occupies the center of the poem, because the most complete version of the royal Park, one including the others, is that of the court and, by extension, the kingdom and the world. By restoring or improving the Park, Charles II exhibits the power of a good king to restore the world.

> Whatever Heaven or high extracted blood
> Could promise or foretell, he will make good;
> Reform these Nations, and improve them more,
> Than this fair Park from what it was before.
>
> (133-36)

The version of the Park as microcosm of the world improved by the King absorbs the other versions, because they are part of the larger world: the gallants and ladies gathering socially, the natural scene (we have observed the symbolic function of the tree imagery), the mythical suggestions (with their emblems of harmony), and the minglings of the natural and human, of art and nature. That larger meaning of the Park as the regal world made into a paradise regained is, in one sense, a meaning led up to with great ease by the sequence of repeated and merging versions of lesser kinds. Such lesser kinds are attractive in themselves, and they need not have contributed to the larger end for this to be an exceptional poem. That they do so is a matter lending tribute to Waller's finer art, his larger capacity. Such capacious art enables us to dwell with firm delight on a new paradise, an improved world, with satisfaction in its individual parts, whether we consider those passages in the poem as we read or, more abstractedly, consider them as themes of the Park as a social gathering place, a lovely natural setting, and a mythic environment. The harmony the poet sought as his theme rises from a harmony of parts that might have been discordant in a lesser poetic music. No doubt darkens my assurance in saying that Waller's poem stands as one of the significant seventeenth-century poems, and one significant precisely after the seventeenth-century manner.

iii. Where Men Abide

The only doubt which I conceive possible is this: why not consider the poem a Restoration public poem? The poem was certainly written during the Restoration and necessarily reflects Restoration experience. But Donne wrote some of his poems even as Spenser was publishing his most important works. The true history involves our saying that the early Donne is an Elizabethan, the late Shakespeare a Jacobean, and Waller in this and other poems a Restoration poet. Certainly, whatever quarrels one has, those with history are the most futile. All I seek are certain discriminations and, at the moment, those involve the social mode. In his *Horatian Ode*, Marvell stands at one with his reader (if his reader could only be sure where Marvell chooses to stand). And in *Astraea Redux*, which frequently traverses paths with Waller's poem, Dryden carries the reader with yet greater assurance (though with less total poetic success). But what to my mind keeps Waller's fine poem social is his reservation. The poet, or whoever we imagine to be the speaking voice of Waller's poem, speaks for himself, refusing to enter into, declining to involve himself with, the scene and the central character of the poem, the King. He eschews direct address and he avoids including

37

the reader in his observations. He says, in effect, this happened, and that is true, and he *implies* consensus rather than asserts it. I feel no doubt but that Waller's stance is one leading to that of Milton speaking to and of Man, or of Dryden speaking of and to men. But the refusal to take in the larger audience, with the willingness to accommodate society by observation of its norms and mores, involves a different poetic procedure from that which was becoming current even as Waller wrote. However sympathetically his poem or Denham's *Cooper's Hill* were viewed in the later seventeenth century, both poets stood off from the poetic scene. In this they curiously resemble that eighteenth-century figure, the Spectator of the world. And it does seem significant that public poetry once more yielded to private in the gradual (and by no means even) transition from poets immersed in the world created by the poems to poets detached from other men. If Waller leads to the full public poetry of Milton and Dryden, the graveyard poets were a step in a detachment that led to Romantic observers willingly or unwillingly cut off from their fellow men. The transition in the seventeenth century involved "the serene contemplator" and "the detached speculator."[21] The social mode of Waller's and other poems laid necessary groundwork for the public mode to evolve. But that is not to presume an identity.

Possibly Jonson or Carew, probably Denham (if he had written more than one outstanding original poem), possibly Marvell might have written *On St. James's Park*. But neither Dryden nor Pope would have written such a couplet as this:

Such various wayes the spacious Alleys lead,
My doubtful Muse knows not what path to tread.

(47-48)

21 My phrases are borrowed from two chapter titles in Maren-Sofie Røstvig, *The Happy Man*, 2 vols. (Oslo and Oxford, 1954), I. iii and v, a study that will prove crucial to my next chapter.

Dryden's Muse, and even more Pope's, felt their uncertainties, but their dubieties reflected other problems. For good or ill, Milton, Dryden, and Pope knew what their worlds were like and shared their knowledge with other men and women. They exhibited an assured sense of their audience, and a familiarity with its claims, that Waller presumed. But they did not have his distance, detachment, and speculation.

> Near this my muse, what most delights her, sees,
> A living Gallery of aged Trees.   (67-68)

Could anyone mistake this for a gesture by Milton, Dryden, or Pope? Waller's setting in a grove and his separation almost from his Muse (who has her own little preferences for trees), contribute to the sense that the poet speaks indirectly for all but immediately for himself. Waller's force derives equally from what he implies he shares with others, and from what he observes himself. Milton and Dryden depend for their bond with the reader on a world they and others fully share. The private world of the Metaphysicals depended crucially on what was not only observed but also valued by the poet himself. The Cavalier poets took the middle road, and if *On St. James's Park* and many other poems gravitated toward the public spectrum, they remained social, refusing to go the whole distance, enjoying a differing integrity.

The important reservations that keep *On St. James's Park* from becoming a public poem have their counterpart in hesitations that insure that the Cavalier poets' more personal poetry will include enough of a shared sense of values to keep a social tone in what I may term their more reserved lyrics. Again, as in Waller's best known poem, the "Song," "Go lovely Rose," we may almost seem to have moved from the Cavalier mode, now to the Metaphysical, although I very much believe that anyone would recognize it as an archetype, or at least as a genuine example, of a Cavalier poem.

Go lovely Rose,
Tell her that wasts her time and me,
That now she knows
When I resemble her to thee,
How sweet and fair she seems to be.

Tell her that's young,
And shuns to have her graces spy'd,
That hadst thou sprung
In desarts, where no men abide,
Thou must have uncommended dy'd.

Small is the worth
Of beauty from the light retir'd;
Bid her come forth,
Suffer her self to be desir'd,
And not blush so to be admir'd.

Then die, that she,
The common fate of all things rare,
May read in thee
How small a part of time they share,
That are so wondrous sweet and fair.

This is not the place to pause for extended discussion of a poem often so lightly read, whether that discussion turns on the traditional matters incorporated with such ease or on the way in which Waller manages to take us so gently to so grim a view. Let us say simply that the situation is something like that in a poem examined earlier, "At Pens-hurst." Now a rose, rather than an estate, is responsive, although the woman remains as aloof as before. The *sic vita* trope of the transient flower is made to mediate beautifully between the two audiences. There is the perfect tact in asking the woman / rose that she "not blush so to be admir'd." And after that good grace (it would be wrong to deem it flattery) there comes at once the harsh address, now to the rose / woman: "Then die." The imperative tells them and us to face reality (again, it would be wrong to deem it a cruelty). "How small a part of time they share": and not only those "That are so wondrous sweet and fair." In "The Sunne Rising," Donne could make a somewhat similar point by saying that when the lovers have come together in full joy, "Nothing else is" (l. 22). Here indeed we find the radical difference between the private and the social mode. To Waller, all that is precious about a rose, or all that makes human *qualities* valuable, is that they are shared socially. There would be no existence of value, "hadst thou sprung / In desarts, where no men abide." To Waller's and the other Cavaliers' experience, even on occasions as personal as one's dismissal by a haughty woman who injures self-esteem, even then the experience was not fully real unless it was confirmed by "men." Contrary to Donne, the world does exist, and everything else is.

No doubt "the social mode" can be described in other terms than those used here to explain how aesthetic distance and poetic handling combine to produce a special angle of vision on the world and oneself. Whatever our terms of reference, the important thing is that we describe and feel a literary phenomenon, an aesthetic possibility realized in practice, a musical key, as it were, of several generations of song. The Metaphysical poets show us how much we may prize the intimate and yet universal exclusiveness of certain kinds of experience. Milton and Dryden, like the Augustans in somewhat different ways, reveal how much we may prize the participating yet personal inclusiveness of certain kinds of experience. What I have termed the social mode and have considered to be the aesthetic radical of Cavalier poetry is that kind of experience that is intimate and yet inclusive, or at times participating and yet exclusive. The social mode, then, held this much in common with the religion for which, and for their estates, the Cavaliers fought: it was a *via media,* a compromise between possible extremes. Like all compromises, it could mean rather differing things to different people subscribing to it, and it could receive now one emphasis or now another, depending on what the situation or the felt need required. If my reader will be good enough to think that all this while I have been particularizing something that he has long since known and felt about poems like "Go lovely Rose," then my modest end has been achieved. We agree that we are talking about a certain kind of poetry with a certain view of life. And we may now turn to ideals of that life, its problems, and its features—in historical reality and in poetic reality.

THE GOOD LIFE

Thus richer then untempted Kings are we,
That asking nothing, nothing need:
Though Lord of all what Seas imbrace; yet he
That wants himself, is poore indeed.
                                            —Lovelace

THE SOCIAL character of Cavalier poetry affords one kind of
testimony to the cohesive civilization behind the work. To
say so much is not to say that all Cavalier poets were alike,
any more than to say that all parts of England were then alike.
More than that, the same men and poets changed with the
times. The question of just what changes took place is one
still exercising social historians. Simply put, however, Cavalier social values are those of an aristocracy and gentry that
two centuries before might have struggled against the throne
but that now sought to protect the King, for all his faults
(Elizabeth, James I, Charles I, Charles II) against his enemies, and to preserve crown, mitre, estates, and what was
often termed "our liberties." But social values often conflicted
with family interests, religious belief, and ambition. There
was much changing of sides, and in fact much of what happened in the century remains beyond our reach. But I see no
need to worry our problems hair by hair or to subtlize them
out of existence. When, on "that memorable Scene," as
Marvell put it, Charles I "with his keener Eye / The Axes
edge did try," he presumably found time to consider that he
had enemies. And those enemies are precisely those who
threw the name "Cavalier" at the King's supporters, as
many of them had thrown "Puritan" at the other side. Not

43

all who supported the King, however, were Cavalier poets. Some men turned to the more intimate alternative of Metaphysical poetry. Very many more were led by piety or indifference to disregard poetry. What deserves remark, therefore, is the variety of those who shared Cavalier poetic and human values, and the degree of agreement among them as to what their values were.

The foremost Cavalier ideal expressed in poetry is what we may call the good life, and in a sense that is the subject of all else following in this book as well as of the preceding chapter. This ideal reflects many things: a conservative outlook, a response to a social threat, classical recollections, love of a very English way of life, and a new blending of old ideas. The ideal is not necessarily Christian or pagan, this worldly or otherworldly. But by comparison with Metaphysical and Restoration poetry, it probably does seem more secular and classical, pagan, or Horatian in some ways than do the other two great modes of seventeenth-century poetry. In itself, Cavalier poetry reveals a consistent urge to define and explore the features of what constituted human happiness, and of which kind of man was good.

One of the best introductions to the idea of the good life may be found in *The Compleat Angler*, whether in Izaak Walton's first part (1653 *et seq.*) or Charles Cotton the Younger's second (1676 *et seq.*).[1] The charm attributed to this work testifies not to a measurability of charm, but to a conviction of decent pleasure that Walton and Cotton carry straight to the heart of even the most indifferent fisherman. Here Walton speaks in the guise of the "angler," Piscator.

I'l now lead you to an honest Ale-house, where we shal find a cleanly room, Lavender in the windowes, and twenty

[1] Walton especially revised his work for reprintings. Some of the best things I quote derive entirely or in their quoted form from later editions. His "naturalness" was both genuine and artistic.

44

Ballads stuck about the wall; there my Hostis (which I may tel you, is both cleanly and conveniently handsome) has drest many a [chub] for me, and shall now dress it after my fashion, and I warrant it good meat. (1, ch. ii; 1653, p. 49)

England is one of the handsomest and pleasantest countries of the world, when the sun shines, and there is a real sense in which the sun shines always in *The Compleat Angler*, even when Peter tells of a rainy day.

Peter . . . indeed we went to a good honest Ale-house, and there we plaid at shovel-board half the day; all the time that it rained we were there, and as merry as they that fish'd, and I am glad we are now with a dry house over our heads, for heark how it rains and blows. Come Hostis, give us more Ale, and our Supper with what haste you may, and when we have sup'd, let us have your Song, *Piscator*, and the [catch] that your Scholer promised us, or else *Coridon* will be dogged . . .

[*Viator*]. And I hope the like [i.e., to be perfect] for my [song], which I have ready too, and therefore lets go merrily to Supper, and then have a gentle touch at singing and drinking; but the last with moderation. (1, ch. xi; 1653, pp. 208-09)

Walton stresses certain elements more heavily, and others more lightly, than a Jonson or a Herrick would; and what he writes turns upon the subject of angling. But his very English book does bring to view a good life, a world in which man's life possesses fullness, and satisfaction is realized.

Walton's small group of anglers exemplifies the usual social microcosm of the Cavaliers. Such another, and more properly poetic, group we find in the famous Tribe of Ben, with Jonson like Piscator the good father figure inculcating virtue, self-knowledge, and poetry (Jonson's mysterious trinity) in his sons, as they express their praise, happiness, and devotion. And their disagreement. One of the wonderful things we discover about Jonson in his relations with his Tribe is his ability to take criticism. His Sons show the complete devotion expected of sons and the frankness expected (in theory, usually) of friends. The evidence, although implicit, drives again and again to show that Jonson expected his Sons to be honest (in not just the modern sense, but also in Walton's and the usual Roman senses). Here is James Howell, writing "To my Father, Mr. Ben Johnson":

> Father *Ben Nullum fit magnum ingenium sine mixtura dementiae*, there's no great Wit without some Mixture of Madness . . . it is verified in you, for I find that you have been oftentimes mad; you were mad when you writ your [*Volpone*], and madder when you writ your *Alchymist*; you were mad when you writ *Catalin*, and stark mad when you writ *Sejanus*; but when you writ your *Epigrams*, and the *Magnetic Lady*, you were not so mad; insomuch that I perceive there be Degrees of Madness in you. Excuse me that I am so free with you.[2]

Free he is, but he spent the next six months hunting out a Welsh grammar for his Father Ben. Jonson's world is, then, a tougher, more verisimilar world than Walton's, so that the goodness and grace that he won for his poetry represent victories far greater than those achieved in Walton's semipastoral world. His tone, his frank, manly, ethical note that led his Sons to respond so forthrightly, can be heard in that poem celebrating the rites of admission to the Tribe.

> Men that are safe, and sure, in all they doe,
> Care not what trials they are put unto;

[2] Howell, *Epistolae Ho-Elianae*, 11th edition (London, 1754), I. v. 16 (27 June 1629).

They meet the fire, the Test, as Martyrs would;
And though Opinion stampe them not, are gold . . .

First give me faith, who know
My selfe a little. I will take you so,
As you have writ your selfe. Now stand, and then,
Sir, you are Sealed of the Tribe of *Ben*.[3]

A true and complete man, inviolate in his central integrity, although passionate enough, Jonson conveys as no other English poet does that sense of *integer vitae*. Reading his ethical poetry, we sense that his central claim (and one that animates us as we read) simply comes to this: "I am a man, and I am true."

Whether or not Jonson's poem just quoted from was addressed to Thomas Randolph, I cannot say, but by what seems a coincidence we have a poem by Randolph offering his gratitude for admittance to the Tribe: "A gratulatory to Mr Ben. Johnson for his adopting of him to be his Son." His gratitude is partly personal and loyal, partly derived from the debt of an artistic "mystery"—whether angling or poetry—that the young owe to their masters.

I was not borne to *Helicon*, nor dare
Presume to thinke my selfe a *Muses* heire.
I have no title to *Parnassus* hill,
Nor any acre of it by the will
Of a dead Ancestour, nor could I bee
Ought but a tenant unto Poëtrie,
But thy Adoption quits me of all feare,
And makes me challenge a childs portion there.
I am a kinne to *Hero's* being thine,
And part of my alliance is divine.

[3] *An Epistle Answering to One That Asked to Be Sealed of the Tribe of Ben*, ll. 1-4, 75-78.

47

Orpheus, Musaeus, Homer too; beside
Thy Brothers by the *Roman* Mothers side;
As *Ovid, Virgil,* and the *Latine Lyre,*
That is so like thy *Horace.*   (1-14)

Randolph's grateful awe conveys part of the good life: its
definition of what is good in terms of relations with contemporaries and with the immortal men of the past. Herrick, addressing Jonson, treats a different aspect of the good life, its
relaxed pleasure (with, he insists, moderation, although no
doubt he and Walton meant different limits by that).

Ah *Ben!*
Say how, or when
Shall we thy Guests
Meet at those *Lyrick* Feasts,
Made at the *Sun,*
The *Dog,* the triple *Tunne?*
Where we such clusters had,
As made us nobly wild, not mad . . .[4]

Whether it is Jonson, Walton, Howell, Randolph, or Herrick—or others—certain elements are shared. Most important
is the premise of the good life, which involves social intercourse with like-minded people. Very frequently that relation
turns on the band of friends, male equals joined in fraternal
affection and esteem. But the relation may be paternal and
filial, as we see in the anglers and the Tribe of Ben. Two
other groups make up the four chief kinds: men and women,
lord and vassal. The latter relation usually involves prince and
subject, but it may be patron and artist. Since all four resemble each other in involving relationship and in representing value, they may be substituted for each other: the woman
loved may be a queen, her lover her subject. In due course

[4] Herrick, "An Ode for him," ll. 1-8.

we shall have cause to observe that such bonds were formed as much by rites and ceremonies as by ethical concerns. We must acknowledge that on occasion some thought that there was just cause for breaking the bonds: Suckling debated the code of love in his lyrics, and most Cavaliers treated the vices of kings and even questions of justifiable regicide in their tragedies. But the usual alignment that imparts Cavalier literature its special flavor is friendship or love or small groups, in each case a relation with a code that implicitly agrees with those in the world whose opinion and principle one values. The limits of the good life and the strength of the social bonds were tested by the Cavalier poets themselves and by events. And we sometimes encounter those moments that make us believe that we are advancing at one step toward pleasure and heaven. A sentence added to later editions of *The Compleat Angler* expresses such anti-Puritanical wisdom in little.

> None of these [trout] have been known to be taken with an Angle, unless it were one that was caught by Sir *George Hastings* (an excellent Angler, now with God).[5]

It is of course the parenthesis that strikes us, a conjunction of many things, but especially of a recreation with divinity, that we may fairly grant to Walton and the Cavaliers as their discovery. Walton's pastoral notes and warmth of heart will be found, with many of the very same details we have noticed, in Cavalier religious poetry. Here is Herrick, in "A Thanksgiving to God, for his House."

> Lord, Thou hast given me a cell
>   Wherein to dwell;
> An little house, whose humble Roof
>   Is weather-proof;

[5] Walton, 1. iv; quoted from the edition of 1750, ed. Browne, p. 49.

Under the sparres of which I lie
    Both soft, and drie . . .
Lord, I confesse too, when I dine,
    The Pulse is Thine,
And all those other Bits, that bee
    There plac'd by Thee;
The Worts, the Purslain, and the Messe
    Of Water-cresse,
Which of Thy kindnesse Thou hast sent; . . .
'Tis thou that crown'st my glittering Hearth
    With guiltlesse mirth;
And giv'st me Wassaile Bowles to drink,
    Spic'd to the brink. . . .
All these, and better Thou dost send
    Me, to this end,
That I should render, for my part,
    A thankfull heart;
Which, fir'd with incense, I resigne,
    As wholly Thine;
But the acceptance, that must be,
    My Christ, by Thee.[6]

The snug house (the fire loved by the Cavaliers comes in a passage omitted), the good food and abundant drink, and the heart, "fir'd with incense," for sacrifice to Christ—all these things make up the good life for which Herrick pays God due thanks. Many will feel that there is more of the *Hesperides* in *The Noble Numbers* than of the latter in the former. There is truth in such a belief. But it is also true that in his secular poems Herrick tends to substitute pagan religious detail for Christian,[7] and that is precisely one of the things

[6] Lines 1-6, 27-33, 37-40, 51-58.
[7] Some of the most valuable specialized studies of Cavalier poets are the essays touching Herrick by Robert H. Deming: "Robert Herrick's

that humanists were about (and for which they were criticized by the more austerely pious). In Herrick as well as in Walton, we discover a lively strain of pleasure and joy of life.

"Sir *George Hastings* (an excellent Angler, now with God)." Walton names a name, adding a title and what is in some sense an occupation. Once again the implication is social: we have our associates, our band of choice, like spirits. And if we seek as before for confirmation in Ben Jonson's poetry, we find him consoling Sir Lucius Cary (later Lord Falkland) for the death of their mutual friend, Sir Henry Morison.

> Call, noble *Lucius*, then for Wine,
> And let thy lookes with gladnesse shine:
> Accept this garland, plant it on thy head,
> And thinke, nay know, thy *Morison*'s not dead.
> Hee leap'd the present age,
> Possest with holy rage,
> To see that bright eternall Day:
> Of which we *Priests*, and *Poëts* say
> Such truths, as we expect for happy men,
> And there he lives with memorie; and *Ben* . . .[8]

Wine and a poetic garland, priests and poets, the named members of a fraternal society—these make up the good life, these provide the marks of "happy men." Unless I am mistaken, Jonson's "happy" is a reverse Latinism of that kind

---

Classical Ceremony," *ELH: A Journal of English Literary History*, xxxiv (1967), 327-48; "Herrick's Funeral Poems," *Studies in English Literature*, ix (1969), 153-67; and his dissertation, "The Classical Ceremonial in the Poetry of Robert Herrick" (Wisconsin, 1965). These studies deal with subjects important in themselves, relevant to other poets, and intimate to some of my concerns here and in ch. iv.

[8] *To the Immortall Memorie, and Friendship of That Noble Paire, Sir Lucius Cary, and Sir H. Morison*, ll. 75-84. *Caveat emptor*: I shall be quoting the passage, and a few others, more than once.

peculiar to him, as when he uses "running judgements" for "current judgements." "Happy" must mean *beati*, and "happy" both in the secular sense and the religious sense of glorification. Above all, Jonson's view presumes a unity of the like-minded and a unity within individual character, *integer vitae*. And for this moral vision, he became the arbiter of his age and "lives with memorie" ever since.

The constituents of such integrity present problems in any discussion. The social elements have been dwelt on in the preceding pages and in the first chapter. Little need be added there. Enough of a paradox remains in a phrase like "an excellent Angler, now with God," however (or in Jonson's call for wine with a vision of eternity), for us to endeavor to distinguish different strains in the *integer vitae*, the good life. I find it convenient to differentiate two strains in the Cavalier conception of the good life, strains distinguishable by the simple expedient of translation into Latin: the *vita bona* and the *vita beata*. And just as the good life may mean the moral life and the happy life, so also may we distinguish the just man, *vir bonus*, and the happy man, *vir beatus*. What marks Cavalier poetry as something of its own kind may partly be the self-awareness of these ideas in the poetry, but perhaps more specially the unusual relation established between the two in poetry. Walton's remark, Herrick's "Thanksgiving," and Jonson's ode reveal the unity of elements that for convenience sake I shall consider separately.

i. *Vita Bona, Vir Bonus*

The moral conception of the good life and the good man of course expresses the ethical idealism of the age. We do not expect the Metaphysical poets to concern themselves with such matters: their poetic minds are "gone out," as Donne puts it, in an ecstasy from our world. We *might*, however, have expected that commentators on the Cavaliers would have concerned themselves with such matters. Since they have not, my role entails saying some fairly obvious things about matters that have not seemed obvious. Obviously, what an ethical system requires is a metaphysics (Christianity provided that) and a social idealism allowing for, or creating, a scheme uniting men and relating them in a society, and allowing an individual his place in it. The hierarchical nature of Tudor and Stuart society certainly provided both relationship and place, even if the terms of provision upset many at the time and would be intolerable today. The rich and various historical study of the early seventeenth century during the past few decades has, unfortunately, brought numerous and often contradictory interpretations. In what follows, I shall not attempt to assess the rise or the fall of the gentry (also, the middle classes, aristocracy, and kingship); prices and wages and monopolies; or the make-up of Parliament and the conflicts of Dissenters and Anglicans and Catholics in the thriddings of York. The Sessions of the Poets were largely conducted on other terms and, sad to say, with very few exceptions, in the seventeenth century the Members of Parnassus were all Conservatives.

And the Cavaliers are Royalists. A central article of their ethical belief entailed a vision of a nation flowing from its spring, the court. Greatness needed to coincide with goodness at the center of the state. The king was the center of authority, and as with any system of government, authority involved legitimacy and power. A glance at what has been called "Cavalier drama" would show how often Cavaliers looked with their eyes and beheld the vices of princes and the corruption of the court in which they served, or to which they gave allegiance. And we often observe Jonson hitting out hard and true against the vices of the court. To be sure, such vices tended to drop from memory when Parliament had won the

Civil Wars, and when Cromwell had come in. But the same ethical system provided the positive values that endured amid the numerous upheavals of the time. Of all the Royalist forms, the masque is première at the Jacobean and Caroline courts. As Carew put it:

> And their faire Fame, like incense hurl'd
> On Altars, hath perfum'd the world.
> Wisdome . . . Truth . . . Pure Adoration . . .
> Concord . . . Rule . . . Cleare Reputation,
> CHORUS
> Crowne this King, this Queene, this Nation.[9]

We should not fail to observe the element of worship and prayer (the imagery resembles that in Herrick's "Thanksgiving"), nor should we fail to observe that the crowning with such virtues is something prayed for, and prayed for on behalf of the nation as much as the King and Queen. Or again, to take the concluding lines of another of Carew's songs:

> Then from your fruitfull race shall flow
> Endless Succession,
> Scepters shall bud, and Lawrels blow
> 'Bout their Immortal Throne.
> CHORUS
> Propitious Starres shall crowne each birth,
> Whilst you rule them, and they the Earth.[10]

Endurance and procreativity, very old human values indeed, inform this Royalist vision. The imagery brings to mind the verses near the close of Marlowe's *Dr. Faustus*:

---

[9] Carew, *Coelum Britannicum*, ll. 1107-1111.

[10] *Ibid.*, ll. 1133-1138; I have omitted speakers' names and changed the italic to Roman usage. Alfred Harbage's *Cavalier Drama* (New York and Oxford, 1936), which one might have hoped to deal with such concerns, is an inducement to further study.

Cut is the branch that might have grown full straight,
And burned is Apollo's laurel-bough.

Carew's vision is triumphant, not tragic, his branch a budding
scepter, his laurel those of earthly and heavenly thrones.

It exceeds the truth to say that the Royalists had all the
good symbols. But we stay within the truth in saying that
those who faced the problem of praising Cromwell and what
he represented used Royalist symbols. Marvell's elegy, *A
Poem upon the Death of O. C.*, has a lengthy passage (ll. 79-
100) that resembles Carew's in imagery, and elsewhere he
seeks to employ the Davidic lore that the Stuarts took especially as their own. How far people went can be judged by
two pictures, one by Vandyke showing Charles I clad in shining black armor on a white horse. The other is the very same
picture, with Cromwell's face painted in.[11] Such efforts no
doubt look like rescue attempts, shifts and evasions, to discover a legitimacy for the rule by a Protector. So they were,
but not only that. They were also expressions of a deep need
to bedeck power in ceremonious state and to crown it with
goodness. As Waller put it praising Cromwell,

> Illustrious acts high Raptures doe infuse,
> And every Conqueror creates a Muse.
> Here in low Strains your milder Deeds we sing,
> But there (my Lord) wee'll Bayes and Olive bring
> To Crown your Head, while you in Triumph ride
> O're vanquish'd Nations, and the Sea beside;
> While all your Neighbor-Princes unto you
> Like *Joseph's* Sheaves pay rev'rence and bow.[12]

[11] See Charles Blitzer, ed., *The Commonwealth of England* (New
York, 1963). At the end is a reproduction of the Vandyke, at the beginning a reproduction of a painting the same in detail, except for
having Cromwell's face.
[12] Waller, *A Panegyric to My Lord Protector*, ll. 181-88.

Marvell and Waller have had arguments put for their political consistency or have been condemned as turncoats, but to my view neither approach satisfies. They did change, and the change was marked. But as Dr. Johnson said of another such occasion, they changed with much of the nation, and with most in that nation they felt a necessity to discover in an individual symbolic of the state those symbols that would reconcile power and goodness, high estate and religion, stability and momentum. From James I, before the revolutions of the century really began, to William III, after they had substantially ended, the needs, and the symbols, of power remained largely the same for King, for regicide Protector, for Protestant (or usurping) Prince, and for poets.[13] And there would also have been agreement with Webster's Antonio at the beginning of *The Duchess of Malfi*, when he said

> a Princes Court
> Is like a common Fountaine, whence should flow
> Pure silver-droppes in generall: But if't chance
> Some curs'd example poyson't neere the head,
> Death, and diseases through the whole land spread.
>
> (I. i. 12-16)

To most of us today, such a passage would seem no more a conception of polity than a basis of the good life. Let us take our concern with royal symbolism a step or two farther, being led by Ben Jonson, to see how the values invested in the King relate to the good man. Good Prince Henry provides a person at once royal, if yet not king, and also an opportunity for Jonson to portray a prince and subject at once in a vision of the *arcanum imperii* with the *vir bonus*.

---

[13] Milton of course used the same royal mysteries and symbols, but positively for God alone. To Royalists the role of Christ as Prophet, Priest, and King shed luster by correspondence on earthy kings, whereas Milton distinguished "Israel's true King" from all others.

You, and your other you, great King and Queene,
Have yet the least of your bright *Fortune* seene,
Which shal rise brighter every houre with *Time*,
And in your pleasure quite forget the crime
Of change; your ages night shall be her noone.
And this yong Knight, that now puts forth so soone
Into the world, shall in your names atchieve
More *ghyrlands* for this state, and shall relieve
Your cares in government; while that yong lord
Shall second him in *Armes*, and shake a sword
And launce against the foes of God and you.[14]

"More *ghyrlands* for this state." The instinct to celebration cannot be repressed. Nor can "the strong Antipathy of Good to Bad," in Pope's phrase: "and shake a sword / And launce against the foes of God and you."

Our next step must be to Jonson himself: man and poet. To try to say it all at once, it is moral integrity that marks the firm center of his poetic compass. Of course we have evidence and to spare showing that Jonson was physically gross, "rocky" as he said, excessive, combative, and (by politer standards) coarse. The evidence also shows that he possessed delicacy, an element of surpassing fancifulness; that he inculcated moderation; that he was a loyal friend; that he possessed a seemingly unerring sense of decorum, beauty, and justice; and that he was admired. Without his massive animal strength, his very virtues might have been insipid. But without his consistent ideals, his integrity, he might have aroused our dislike.

What a poet praises (or criticizes) in others usually tells us something of his own ideals. When Jonson acclaimed someone for being "always himselfe and even," we see very clearly the ideal of *integer vitae*. In a few lines already quoted,

---

[14] Jonson, *Prince Henries Barriers*, ll. 419-29.

we see that self-knowledge is as essential to Jonson as to Socrates.

> First give me faith, who know
> My selfe a little. I will take you so,
> As you have writ your selfe.[15]

Knowing oneself and others implies a capacity to trust oneself and others. It also implies that one will remain oneself with constancy in a world of threat and change.

> Well, with mine owne fraile Pitcher, what to doe
> I have decreed; keepe it from waves, and presse;
>   Lest it be justled, crack'd, made nought, or lesse:
> Live to the point I will, for which I am man,
>   And dwell as in my Center, as I can.[16]

The traditional emblems are here: the vessel of the soul, the threatening waves, the center for constancy, and the circle for perfection. And what such emblems help portray is precisely a good man defining his place in the world.

In that world, and as part of the moral landscape of Jonson's poetry, there rise shapes and forces threatening the constant man. As Milton said, we are no longer capable of defining good without a knowledge of evil, and Jonson often provides ample shadowing for his bright ideals. Even his panegyrics bear warnings for princes and kingdoms.

> Who, *Virtue*, can thy power forget,
> That sees these live, and triumph yet?
> Th' *Assyrian* pompe, the *Persian* pride,
> *Greekes* glory, and the *Romanes* dy'de:
>   And who yet imitate
> Theyr noyses, tary the same fate.
>   Force Greatnesse, all the glorious wayes

---

[15] Jonson, *An Epistle . . . the Tribe of Ben*, ll. 75-77.
[16] *Ibid.*, ll. 56-60.

You can, it soon decayes;
But so *good Fame* shall, never:
Her triumphs, as theyr Causes, are for ever.[17]

Merlin gives to Chivalry a Jonsonian caution in matters of
praising kings, in a passage coming at its end to the high ideal
of integrity which Jonson set.

Nay, stay your valure, 'tis a wisdome high
In Princes to use fortune reverently.
He that in deedes of *Armes* obeyes his blood
Doth often tempt his destinie beyond good.
Looke on this throne, and in his temper view
The light of all that must have grace in you:
His equall *Justice*, upright *Fortitude*
And settled *Prudence*, with that *Peace* indued
Of face, as minde, always himselfe and even.
So HERCULES, and good men beare up *heaven*.[18]

The legend of Hercules relieving Atlas of his burden of the
world worked throughout the century as a political type for
kings.[19] What sets off Jonson's use of the type is the domi-
nance of political and other concerns by the moral. To me,
the moral element lives in Jonson's very cadence and syntax.
To others, I can point to the cautioning of princes to moderate courses, the insistence on integrity ("always himselfe
and even"), and the equation of "good men" with the Hercules / king typology. The regal is assimilated in the individual,
in the very manhood of the good man.

The good life finds definition, then, in terms of other lives,
whether those of kings or of commoners. But the act of de-

---

[17] Jonson, *The Masque of Queens*, ll. 764-73.
[18] Jonson, *Prince Henries Barriers*, 405-14.
[19] In *Mystagogus Poeticus*, p. 116, Ross writes that Hercules can be
"The type of a good king"; for citation and further lore, see ch. i, n. 15,
above.

fining involves the discovery of self-knowledge, and in order to be defined as a good man one must be constant. On assumptions such as these, Jonson creates an artistic fiction no less imaginative than that of the writer of love lyrics. And his fiction enables him to laugh at himself, to drop the adventitious in human life, in order to stress what is central to goodness. As he conceded in addressing Sir William Burlase ("My Answer. The Poet to the Painter"):

Why? though I seeme of a prodigious wast,
I am not so voluminous, and vast
But there are lines, wherewith I might b'embrac'd. . . .

But, you are he can paint; I can but write:
A Poet hath no more but black and white,
Ne knowes he flatt'ring Colours, or false light.

Yet when of friendship I would draw the face,
A letter'd mind, and a large heart would place
To all posteritie; I will write *Burlase*.   (1-3, 19-24)

He draws himself good humoredly as a gross man, so that what he says about himself as writer and about Burlase is accepted as truth. (Jonson also plays with the two "sister arts," of course.) By such means he creates a fiction of himself, his friends, and his King as good men. It must be stressed that that fiction is as much something made, or rather, as much a transformation of reality, as anything in Donne. But like every good fiction, it carries the stamp of artistic currency within the perfect ring of its own coinage. All these considerations join in Jonson's self-awareness.

Let me be what I am, as *Virgil* cold,
    As *Horace* fat; or as *Anacreon* old;
No Poets verses yet did ever move,
    Whose Readers did not thinke he was in love.[20]

[20] Jonson, *An Elegie*, ll. 1-4.

The conviction is necessary. Jonson convinces us that, like his Prince, James I, he was a good judge of good men. More than that, and unlike James, he also had good judgment of bad men and of himself. And if the portrait I have been getting Jonson to draw of himself makes him seem too bluff, let me refer to the word "thinke" in the last line quoted, or ask if there is not some fun in the first two lines, and wonder if it might not be worthwhile to be cold, fat, and old if one could also be an English poet combining the geniuses of Virgil, Horace, and Anacreon? My aim has been to emphasize the strength of Jonson's moral architecture, because that is, to my view, the essential, the functional aspect of his art.

The ideal of the good life finds its way into all manner of poetic forms. We have seen it in the masque, the ode, poems of address, and lyrics. As the last chapter of this study will show, poems of address to male friends express this concern particularly well. But we find it in epigrams, epitaphs, elegies, epistles, songs, and other forms. What is usual, however, is some element of address, whether at the beginning or end, or whether fictional altogether, or whether systematic. Not infrequently the titles, if in English, read, "To my worthy Friend," "On a . . . ," "To one that . . ."; or if in Latin, "Ad . . ." or "In . . ." In other words, the good life tends to be defined in terms of good lives; the good life finds understanding in terms of individual good men. Having defined our terms with the assistance of Jonson, we may turn to Lovelace for help in seeing how the terms may work.

It should prove useful to begin with the unfamiliar, to take a poem that is not a household word: "To my Worthy Friend Mr. Peter Lilly: on that excellent Picture of his Majesty, and the Duke of Yorke."

See! what a *clouded Majesty*! and eyes
Whose glory through their mist doth brighter rise!

See! what an humble bravery doth shine,
And griefe triumphant breaking through each line . . .
    These [sitters] my best *Lilly* with so bold a spirit
And soft a grace, as if thou didst inherit
For that time all their greatnesse, and didst draw
With those brave eyes your *Royall Sitters* saw. . . .
So that th'amazed world shall henceforth finde
None but my *Lilly* ever drew a *Minde.*

(1-4, 17-20, 31-32)

This is not a great poem, but it does have a vision of human worth. The episode it treats is that of Charles I and the Prince of Wales held at Hampton Court in the charge of the Duke of Northumberland. Lely bravely undertook to paint the King in his *"clouded Majesty,"* and to express the King bearing up nobly in his misfortunes. Lovelace seeks to catch what must be called Lely's loyal art, or art of loyalty, and the praise he gives to the painter is praise of which he is worthy himself as a poet.

A yet gloomier situation lies behind "To his Deare Brother Colonel F. L. immoderately mourning my Brothers untimely Death at Carmarthen." The poet and eldest son consoles the second son, Colonel Francis Lovelace, under whose command a younger brother (probably William) had been slain in Welsh action during the First Civil War. Explaining that Fate whips us like children first to make us weep, and then *because* we weep, he concludes:

IV

Then from thy firme selfe never swerve;
Teares fat the Griefe that they should sterve;
I'ron decrees of Destinie
Are ner'e wipe't out with a wet Eye.

62

V

> But this way you may gaine the field,
> Oppose but sorrow and 'twill yield;
> One gallant thorough-made Resolve
> Doth *Starry Influence* dissolve.

It was a real war, and real people were killed. The idealizing of the court by the Cavaliers no doubt seems remote from modern experience, but setting side by side the troubles of Charles I and the Lovelace family, we can bring ourselves somewhat closer to the seventeenth century. There is no need to sentimentalize the Cavaliers. Their enemies gave them the name to suggest that they were swaggerers, abusers, and dissolutes, as not a few of them no doubt were. Lovelace's very name might be held against him. But there is no need to be cynical about the Cavaliers. It is enough to say that Lovelace very well reveals that when the Cavaliers fell on hard times they found strength within, they proved their own moral resources. Later we shall have occasion to see how frequent the movement to within marks their poetry; but now it is the ethos of a class of men and a class of poetry that concerns us. They were sustained partly by the sense of class and tradition, but also by the conviction of acting in a just cause. As Sir Philip Warwick, secretary to Charles I, wrote after the royal cause had gone down in defeat, "When I think of dying, it is one of my comforts, that when I part from the dunghill of this world, I shall meet . . . King Charles and all those faithful spirits that had virtue enough to be true to him, the Church and the laws, unto the last."[21] The Cavaliers had *their* Good Old Cause, and lost it, before the Puritans had and lost theirs.

We may return at this juncture for another look at Love-

[21] Quoted by Margaret Barnard Pickel, *Charles I as Patron of Poetry and Drama* (London, 1936), p. 19.

63

lace's poem most familiar to our generation, "The Grasse-hopper."[22] The poem begins with six stanzas addressed to that Anacreontic creature, which has "The Joyes of Earth and Ayre . . . intire" (5), at least until "the Sickle" and winter come. The "Joyes" of the insect "Bid us lay in" against the cold season and to "poize" against the floods of state "an o'reflowing glasse." The rites of friendship by the fireside will preserve us. In the first stanza of address to Charles Cotton the Elder, "Thou best of *Men* and *Friends*," Lovelace opposes friendship to fate, but such triumphant opposition is possible only when the friend is the "best of *Men*." The *vir bonus* is central to friendship and the rites celebrated in the Cavalier winter.

The eighth stanza provides a tableau in some ways too difficult to sort out. King December will soon have "his Raigne" (with a pun on "rain") usurped, but the "show'rs" of wine drunk by the friends will restore it. December also stands for Christmas. In 1647 there were disturbances throughout England over the Puritans' closing the churches, opening the shops, and banning the traditional festivities.[23] To the Puritans, Christmas and its revelries were as corrupt and papistical as the stained glass that they succeeded almost totally in destroying in English churches. To have "Christmas" officially banned as a word and a feast meant important if different things to Puritans and Cavaliers alike. So much comes easily from the poem and history. In what follows I shall have to strain the fabric, because Lovelace both implies and (to my mind) also withholds a political significance. Without saying

[22] For the text of the poem, see the Appendix.
[23] See Samuel L. Gardner, *History of the Great Civil War*, 3 vols. (London, 1886-1891), III. 281-82. On 3 June 1647 was issued the Parliamentary ordinance forbidding traditional celebration of Christmas, Easter, and Whit Sunday. Evelyn records that Puritan soldiers raided an Anglican church service on Christmas Day, 1657. One may compare with all this Jonson, *Christmas his Masque* (1616).

so much, he suggests that the role of Christ the King, and King of Kings, and by correspondence the earthly king, will return to their reigns again. Such correspondence between religion and politics marks the century, as Warwick's remarks show, or as does the famous "No King, no Bishops" of James I. Some of the further reaches of the poem are yet more difficult, and I shall but touch on them here, deferring extended discussion to the last chapter. The drinking of the good friends (to Christmas and King Charles) during the winter corrects, without wholly contradicting, the example of the grasshopper's summer carousal. The insect is after all a "Poore verdant foole" (which is not the worst description of Charles I), whereas the friends create the "Genuine Summer" in each other. The last stanza takes us to the self-sufficiency of the good man.

> Thus richer than untempted Kings are we,
>   That asking nothing, nothing need:
> Though Lord of all what Seas imbrace; yet he
>   That wants himselfe, is poore indeed.

No doubt Lovelace's conclusions about the ceremonies of wine as a defense measure will suggest the extent to which the morally good life and the enjoyably good life tend to merge. Some reasons for that tendency to grow will be given in the next section of this chapter. In the rest of this part, my aim is, by returning to Jonson, to define some characteristics not only of the good life but also of the ethical vision created by Cavalier poetry. Jonson's "Epistle to Sir Edward Sacvile, Now Earle of Dorset" shows how central to the ethical vision is self-knowledge.

>   let them then goe,
> I have the lyst of mine owne faults to know,
> Looke [to] and cure; Hee's not a man hath none,

> But like to be, that every day mends one,
> And feels it . . .   (113-17)

One must look inward, improve upon oneself: and *feel* the process of betterment. Such a good man is not the ontological entity favored by many today: being is not enough, and time as well as action are necessarily involved. A bad man does not become Sir Philip Sidney in a day, nor does a fool like Tom Coryat become wise overnight.

> Men have beene great, but never good by chance,
> Or on the sudden. It were strange that he
> Who was this Morning such a one, should be
> *Sydney* e're night? or that did goe to bed
> *Coriat*, should rise the most sufficient head
> Of Christendome?   (124-29)

Or, more emphatically,

> Yet we must more then move still, or goe on,
> We must accomplish; 'Tis the last Key-stone
> That makes the Arch.   (135-37)

How appropriate it seems that Jonson gave up the brick-layer's trowel in order, years later, to create a moral architecture so fine as this. These passages, it may be added, are worlds removed from "Drinke to me, onely, with thine eyes," not just in thought but in the very voice and tone. The world that we are considering is the world of ethical issues, and that world is built and sustained by elements other than ideas alone. Jonson's ethical verse speaks out in lines as "strong" as those in much of the poetry of Donne or Chapman. The lines are heavily cut by caesuras; they spill over one to the next to the next; and the words, even the syntax, crowd together with what seems a plain-spoken indifference to fine phrasing.

Jonson's finest achievement probably lies in his ability to

arouse in us a total conviction of his moral integrity, of the good man searching himself and others, of the good life being glimpsed and sought for with strenuous effort. And it does not exclude other qualities. When he addressed various of his contemporaries, his skill with epigram frequently produced wonderful *obiter dicta*.

And, in those outward formes, all fooles are wise. . . .
For he, that once is good, is ever great.[24]

*

Truth, and the Graces best, when naked are.[25]

*

Who can behold their Manners, and not clowd-
Like upon them lighten? . . .
That scratching now's our best Felicitie? . . .
Informers, Masters both of Arts and lies . . .
These take, and now goe seeke thy peace in Warre,
Who falls for love of God, shall rise a Starre.[26]

The tone varies considerably in such lines, but all contribute to a sense of truth, of integrity, and of responsible moral judgment. The ideas matter, and so does the style, which has been called "plain,"[27] and which certainly is created to give the effect of simple moral integrity.

Jonson's conception of the good man is precisely that, a conception, an artistic creation and a view of life. How far we may go toward identifying the values in what is created with those in the life is always a question. Physicians often

[24] Jonson, *Epistle to Katherine, Lady Aubigny*, ll. 36, 52.
[25] Jonson, *An Epistle to Master John Selden*, l. 4.
[26] Jonson, *An Epistle to a Friend, to Perswade Him to the Warres*, ll. 60-61, 144, 164, 195-96. This is one of Jonson's finest and bleakest poems.
[27] See the very learned study by Wesley Trimpi, *Ben Jonson's Poems, A Study of the Plain Style* (Stanford, 1962), who is especially helpful in adducing classical material. My crucial stress upon Jonson's plain-spoken ethical truth will suggest my indebtedness to Trimpi.

neglect their health or mistreat themselves, and the moralists often find it easier to lead than to follow. We have the evidence of the extraordinary number of poems celebrating him after his death to understand that although his contemporaries knew Jonson's considerable faults ("Hee's not a man hath none"), they accepted them and the man because of his far greater virtues. And although some may question whether poetic evidence can be accepted for biographical fact, this much is certain: Jonson had the ability as a poet (and *I* think the fortitude as a man) to meet disaster. The epigrams on his son and daughter are too well known to quote. A less familiar poem, and one of Jonson's most charming, humorously remonstrates with the god of fire for burning his house in November, 1623. *An Execration upon Vulcan* begins:

> And why to me this, thou lame Lord of fire,
>   What had I done that might call on thine ire?
> Or urge thy Greedie flame, thus to devoure
>   So many my Yeares-labours in an houre?
> I ne're attempted, *Vulcan*, 'gainst thy life;
>   Nor made least line of love to thy loose Wife.

So much for the smithy and his wife, the drab of Mars. But Jonson also touches on a terrible personal loss, nothing less than his books and a body of works in manuscript. Or rather,

> I dare not say a body, but some parts
>   There were of search, and mastry in the Arts.
> All the old *Venusine*, in *Poëtrie*,
>   and lighted by the *Stagerite*, could spie,
>   Was there made English . . .   (87-91)

Not only did he lose Horace fortified by Aristotle, but also an English grammar, a translation of Barclay's romance, *Argenis* (at royal request), and history,

Wherein was oyle, beside the succour spent,
  Which noble *Carew, Cotton, Selden* lent:
And twice-twelve-yeares stor'd up humanitie,
  With humbler Gleanings in Divinitie,
After the Fathers, and those wiser Guides
  Whom Faction had not drawne to studie sides.

(99-104)

The equanimity seems very Stoical, and in fact he gives more credit to his sources and helpers than to himself. The wit at times enjoys itself, as he thinks of Vulcan's arson.

Well-fare the Wise-men yet, on the *Banckside,*
  My friends, the Watermen!   (123-24)

Even if they were of no help and are wise only in that area! Or again, "Pox on your flameship, *Vulcan* . . ." (191). And at the end the personal loss is rendered into a vision of what men most seek. For it would have been all right in warlike areas such as the Netherlands to

Blow up, and ruine, myne, and countermyne,
  Make your Petards, and Granats, all your fine
Engines of Murder, and receive the praise
  Of massacring Man-kind so many wayes.
We aske your absence here, we all love peace.

(205-209)

We shall see repeatedly how widely the desire for peace was shared by Cavalier poets, and how central it was to very different views of the good life. But surely we have also seen a poet who proved a man in adversity.

The good man has one further function besides knowing himself, improving himself, and conquering his passions with equanimity. He must judge good and evil in other men. Many of Jonson's poems judge evil through the resources of satire, and others judge good through panegyric. How conscious

Jonson was of the role of the good man as judge can be seen from his praise of one of the most learned men of the day in "An Epistle to Master John Selden."

> What fables have you vext! what truth redeem'd!
> Antiquities search'd! Opinions dis-esteem'd!
> Impostures branded! And Authorities urg'd!
> What blots and errours, have you watch'd and purg'd
> Records, and Authors of! how rectified,
> Times, manners, customes! (39-44)

Jonson possessed many poetic and human virtues, and some of them are at least implied there. He prized learning that brought moral experience, wisdom. He may cleverly adapt the old proverb that a wise man is at home in all countries, including Selden among those "that have beene /Ever at home: yet, have all Countries seene" (29-30). But he does not stop there, adding that his friend is

> . . . like a Compasse keeping one foot still
> Upon your Center, doe your Circle fill
> Of generall knowledge; watch'd men, manners too,
> Heard what times past have said, seene what ours doe. (31-34)

A quiet wit voices itself in the distinction between hearing what is remote and seeing what is at hand. It is as though Selden has talked with the wise men of the past. The compass image conveys a virtue important to Jonson, constancy,[28] and the circle emblem of course perfection.

These many aspects of the Cavalier conception of *vir bonus*

---

[28] See Rosemary Freeman, *English Emblem Books* (London, 1948); there is a very full discussion of lore associated with compasses in John Freccero's essay, "Donne's 'Valediction Forbidding Mourning,'" *ELH: A Journal of English Literary History*, xxx (1963), 335-76, which touches on Jonson at n. 43. Since Jonson took for himself the emblem of the broken compass, we see that he possessed humility, after all. In my view, that image is appropriate for the world of the plays but not for the poet's vision of that world.

and of *vita bona* seem all to come together in Jonson's regular pindaric ode, *To the Immortall Memorie, and Friendship of That Noble Paire, Sir Lucius Cary, and Sir H. Morison.* As the poem tells us, Sir Henry Morison died (*aet.* about twenty) in battle. Sir Lucius Cary, later Viscount Falkland, also later to fall in battle, was as much an ideal courtier as the legendary Sir Philip Sidney (who also died in the field). Falkland married Morison's sister and styled himself a "brother" of Ben Jonson. Such adoption of relationships enables Jonson to write with great feeling and integrity about his subject, which, in its initial stages, consists of the negative contrast (1-42) and then the positive value represented by Morison.

> Hee stood, a Souldier to the last right end,
> A perfect Patriot, and a noble friend,
> But most a vertuous Sonne.
> All Offices were done
> By him, so ample, full, and round,
> In weight, in measure, number sound,
> As though his age imperfect might appeare,
> His life was of Humanitie the Spheare.   (45-52)

Like the circle, the sphere is an emblem of perfection and here is suggestive of the macrocosm. Words less obvious in import provide the crucial, distinctive features of Jonson's vision of the good man, however. "Hee stood"—the simple declaration implies constancy and self-sufficiency in a world of strife, *mens immota manet.* How meaningful this seemingly passive or defensive posture may seem to a world with values other than our own can best be judged by recalling the climax of *Paradise Regained*:

> To whom thus Jesus. Also it is written,
> Tempt not the Lord thy God; he said and stood.
> But Satan smitten with amazement fell.   (IV, 560-62)

71

Repeatedly in the seventeenth century the moral crisis centers on standing or falling, or, in another version, of fleeing from the doomed city: Whither shall I fly? What shall I do to be saved? "Hee stood." That this does not mean merely passive or defensive action can be shown by what follows: Morison performed "All Offices." This, the central word of the passage, derives from the Stoic word for obligations or duties, and to anyone in the century who could understand such things it would immediately call to mind "Tully's *Offices*," the very popular *De Officiis* by Cicero.[29]

We may recall, "For he, that once is good, is ever great," in reviewing the best known stanza of the poem, which now contrasts falling with another image.

> It is not growing like a tree
> In bulke, doth make a man better bee;
> Or standing long an Oake; three hundred yeare,
> To fall a logge, at last, dry, bald, and seare:
> A Lillie of a Day,
> Is fairer farre, in May,
> Although it fall, and die that night;
> It was the Plant, and flowre of light.
> In small proportions, we just beauties see:
> And in short measures, life may perfect bee.
>
> (65-74)

The "flowre of light," that supererogatory image, runs through numerous versions in the next three stanzas, culminating in "Two names of friendship, but one Starre: / Of hearts the union" (98-99). The old trope for friendship, one heart in bodies twain, is altered here into the stellar apotheosis that emerges in Jonson's handling of the light imagery radiating from the lily. But the stanza's formal properties also deserve attention. In the central two lines (the fifth and

[29] The concluding pages of this chapter will discuss the relevance of Cicero and Stoicism to Cavalier conceptions of the good life.

sixth) we see the "Lillie of a Day," and these two short lines represent the "small proportions," "just beauties," and "short measures" that Jonson speaks of in the last two lines. More than that, there is a circularity of rhyme scheme, with the rhyme sounds of the first and last couplets being the same. The circularity is complete in the shift from "better bee" (66) to "perfect bee" (74). The stanza almost seems an unspoken version of Jonson's compass conceit.

The elegiac reconciliation completes itself in such perfection, so that what remains is the celebration we have partly read on earlier pages.

> Call, noble *Lucius*, then for Wine,
> And let thy lookes with gladnesse shine:
> Accept this garland, plant it on thy head,
> And thinke, nay know, thy *Morison*'s not dead.
> Hee leap'd the present age,
> Possest with holy rage,
> To see that bright eternall Day:
> Of which we *Priests*, and *Poëts* say
> Such truths, as we expect for happy men,
> And there he lives with memorie; and *Ben*
>
> *Johnson*, who sung this of him, e're he went
> Himselfe to rest,
> Or taste a part of that full joy he meant
> To have exprest,
> In this bright *Asterisme*.   (75-89)

The imagistic development from the "flowre of light" must strike every reader, and so too the introduction of the poet's name. In fact, seventeenth-century elegies from Donne's *Anniversaries* to Dryden's *Eleonora* usually do not allow names, unless pastoral appellations, in the text of the poem. Jonson is not so "classical," but he does employ the oratorical device of "ethical proof" or personal testimony in that remarkable division of his name between two stanzas. Toward the end of the second of these stanzas, Jonson speaks of Falkland's "*Harry*," that is, Sir Henry. At the beginning of the joined stanzas there appears "noble *Lucius*," at the end "his *Harry*." What bridges the two is "*Ben / Johnson*"—the man who identifies himself as priest and poet saying "Such truths, as we expect for happy men" (82-83).

Without a forfeit of social relationships, the sense of personality runs throughout the poem, and indeed throughout Jonson's poetry. No shrinking violet, he extolls himself among the flowers of light. The conviction—and the imagistic resources enabling the reader to relive the conviction—of the good man and the good life provide the firm ethical center of Cavalier poetry. Loyalty to the King, and to "the Church and the laws," as Sir Philip Warwick put it, provided Cavalier poetry with the social basis for its integrity. In a nearer concentric of that sphere there lay ideals of integrity, constancy, self-sufficiency and, if need be, self-sacrifice. The Cavaliers could not make exclusive claim to such ideals, any more than could the Puritans to faith and service in a providential cause. In fact, Falkland later went off to the wars with divided mind and with what has seemed to many a deliberate, almost suicidal excess of bravery. Such men did not return from their Battles of Newbury: Falkland was not a friend whom one needed "to Perswade . . . to the Warres." Whether to stand, to fly to death in what one knew or hoped to be a good cause, or to brighten like the "flowre of light," it was important that one be oneself and be good. As Jonson's learned friend Selden put it concerning books,

> I would call *Books* onely those which have in them either of the two objects of Mans best part, *Verum* or *Bonum*, and to an instructing purpose handled.[30]

[30] Selden, *Titles of Honor* (London, 1614), sig. *a2*ʳ⁻ᵛ.

We recognize Jonson's tone in those words, and even more wholly in a distinction Selden draws at length.

> So Generous, so Ingenuous [sincere], so proportion'd to good, such Fosterers of Vertue, so Industrious, of such Mould are the *Few*: so Inhuman, so Blind, so Dissembling, so Vain, so justly Nothing, but what's Ill disposition, are the *Most*.[31]

And it was of Jonson, to whom the Cavaliers owed their rich ethical patrimony, that Selden could write: that is, of

> my beloved friend that singular Poet M. *Ben*: *Jonson*, whose speciall Worth in Literature, accurat Judgement, and Performance, known only to that *Few* which are truly able to know him, hath had from me, ever since I began to learn, an increasing admiration.[32]

Sir John Beaumont caught the accent in his poem on Jonson's death, saying of him,

> Since then, he made our Language pure and good,
> To teach us speake, but what we understood . . .
> And though He in a blinder age could change
> Faults to perfections, yet 'twas farre more strange
> To see (how ever times, and fashions frame)
> His wit and language still remaine the same
> In all mens mouths . . .
> Could I have spoken in his language too,
> I had not said so much, as now I doe,
> To whose cleare memory, I this tribute send
> Who Dead's my wonder, Living was my Friend.[33]

[31] *Ibid.*, sigs. *a*2ᵛ-*a*3ʳ.
[32] *Ibid.*, sig. *d*1ʳ; I have reversed roman and italic usage.
[33] "To the Memory of him who can never be forgotten, Master Benjamin Johnson," ll. 41 ff. from the collection made by Falkland, *Jonsonus Virbius*; in *Ben Jonson*, ed. C. H. Herford, Percy and Evelyn Simpson, 11 vols. (Oxford, 1925-1952), XI, 438-39.

Jonson would have taken pride in that identification with himself and goodness and right language, just as he would have recognized the imagery and the friendship of Falkland's poem in *Jonsonus Virbius.*

> I then but aske fit Time to smooth my Layes,
> (And imitate in this the Pen I praise)
> Which by the Subjects Power embalm'd, may last,
> Whilst the Sun Light, the Earth doth shadows cast,
> And feather'd by those Wings fly among men,
> Farre as the Fame of Poetry and BEN.[34]

O Rare Ben Jonson! Neither the first nor the last English poet to think that his poetry was a moral as well as a beautiful art, he alone put the good man at the center of his poetry and made us believe that he was such, for all his follies. His imagination was fanciful enough to create new forms for the stage or to present to him as he lay abed the Romans and Carthaginians doing battle about his great toe. His humanity was such that he grieved as a father for the loss of son and daughter and yet could adopt others to his Tribe. But with his learning it was his conviction of the centrality of the good life to poetry that won him the unparalleled respect of his contemporaries.

ii. *Vita Beata, Vir Beatus*

The blessed or happy life of the Cavaliers is of course not beatific in the usual religious sense, since religious matters are subsumed in the *vita bona.* The happy life is good (*beata*) because it possesses, or may possess, enjoyments and a fullness of comfort. We believe that the happy life was known by Walton's group of anglers with their pots of ale, singing before a fire while their hostess cooks the fish carefully as in-

---

[34] *An Eglogue,* ll. 285-90; *Ben Jonson,* xi, 437.

structed after Master Walton's way, snug together while the
rain falls, certain in the knowledge that their rooms have
clean sheets and "Lavender." There is condescension in
Pope's witty paradox on "The Mob of Gentlemen who wrote
with Ease" and exaggeration as well; but there is also envy. No
such group could be imagined at the court of the Georges
and Walpole. By the end of the seventeenth century, Dryden
could look back on the first four decades with a nostalgia for
a merry England, a life of country gentry that seemed to have
a vanished innocence. "Enter *Diana*."

> With Horns and with Hounds I waken the Day,
> And hye to my Woodland walks away;
> I tuck up my Robe, and am buskin'd soon,
> And tye to my Forehead a wexing Moon.
> I course the fleet Stagg, unkennel the Fox,
> And chase the wild Goats o're summits of Rocks,
> With shouting and hooting we pierce thro' the Sky;
> And Eccho turns Hunter, and doubles the Cry.

And the part of the chorus appropriate to this section of *The
Secular Masque* sums up what the happy life looked like in
retrospect.

> *Then our Age was in it's Prime,*
> *Free from Rage, and free from Crime,*
> *A very Merry, Dancing, Drinking,*
> *Laughing, Quaffing, and unthinking Time.*

This is of course too simple. England was always merry—two
generations before—and Dryden is investing his country boy-
hood with a happiness that he had surely not then known. But
his vision probably owes something as well to the writings of
Jonson and his Sons.

It will be recalled that in his "Answer" to *Gondibert*,
Thomas Hobbes distinguished the genres of poetry according

to narrative or dramatic treatment of three worlds: court, city, and country.[35] Characteristically, though not inevitably, the good life (*vita bona*) was one led or earned amid the dangers at the court or in the city. The happy life (*vita beata*) was in like wise most readily found in the country, where a few friends, at least one friend and up to four or five or so, gathered and made their own world. But having dwelt on Jonson and his friends, I prefer now to choose to represent the happy life in another of its microcosmic terms as a life welcoming women and love. We may for the moment restrict our view of the happy life yet further by attending to the Cavaliers' seduction poems and other amatory verse. We can remind ourselves later of the restricted nature of our present view.

The two poets regarded by their contemporaries or ours as most typical of the hedonistic Cavalier are no doubt Thomas Carew and Sir John Suckling. Carew has interested modern readers the more, partly because his poems have long since been handsomely printed in a reliable edition, and also because his Cavalier songs often bear a Metaphysical accent.[36] To his contemporaries, Carew (like Suckling) was a poet of "Witt," who lived the kind of life depicted in Donne's *Songs and Sonnets* and who, it was said, "speaks raptures."[37] The remark touches specifically on Carew's then most popular, and

[35] Hobbes, "Answer," in *Critical Essays of the Seventeenth Century*, ed. J. E. Spingarn, 3 vols. (Bloomington, 1963), II, 54-55. It is less well known that *Gondibert* itself is an attempt to define the worlds of court, city, and country.

[36] Carew's splendid elegy on Donne is well known, as is also the fact that it is modeled to some extent on Jonson's elegy on Shakespeare. But Carew's lines (11-12) of the poem, *In Answer of an Elegiacall Letter*, seem not to be well known: a proper poem could not be written on the subject by "*Virgil*, nor *Lucan*, no, nor *Tasso* more / Then both, not *Donne*, worth all that went before."

[37] See "Carew's Early Reputation," Rhodes Dunlap's Introduction in *The Poems of Thomas Carew* (Oxford, 1957), p. xlvi.

highly erotic, poem, A *Rapture*, and perhaps on "The second
Rapture" (a negligible piece). The raptures are those that
Milton's Comus proffers to the Lady. Indeed it almost seems
that Comus had been reading Carew's "To A. L. Perswasions
to Love," telling her not to be "proud."

> For being so, you loose the pleasure
> Of being faire, since that rich treasure
> Of rare beauty, and sweet feature
> Was bestow'd on you by nature
> To be enjoy'd, and 'twere a sinne,
> There to be scarce, where shee hath bin
> So prodigall of her best graces.   (7-13)

A crucial verb here (and in l. 15) is "enjoy." No small part
of the verse addressed to women by Cavaliers is poetry of
enjoyment in such senses. Such poems are, as another of
Carew's titles so elegantly puts it "Perswasions to enjoy." In
"To A. L.," he seems to ask womankind to join the race and
make it happy.

> Were men so fram'd as they alone
> Reap'd all the pleasure, women none,
> Then had you reason to be scant;
> But 'twere a madnesse not to grant
> That which affords (if you consent)
> To you the giver, more content
> Then me the beggar.   (19-25)

"Content" and "contentation" present other central terms
echoing from Carew to Walton and Cotton. And there is not
a little of Marvell's style in those lines—topics, meter, and
the logic of conditionals or subjunctives followed by declaration. But Carew has a more nervous style that seeks to avoid
rather than to suspend contraries.

Carew's vision of "enjoyment" is distinctive by virtue of

the participation of the woman, by a fuller sense of "rapture" (as opposed to arousal) than elsewhere in seventeenth-century poetry, and by a tendency to idealize rapture by pagan myth. A simple example may be found in "Upon a Mole in Celia's bosome."

> That lovely spot which thou dost see
> In Celias bosome was a Bee,
> Who built her amorous spicy nest
> I'th Hyblas of her either breast. (1-4)

So great is the sweetness there that the bee chokes and expires.

> Yet still her shaddow there remaines
> Confind to those Elisian plaines;
> With this strict Law, that who shall lay
> His bold lips on that milky way,
> The sweet, and smart, from thence shall bring
> Of the Bees Honey, and her sting. (15-20)

We can best announce the country of the mind we have come to by paraphrasing the remark of the Sea Captain in *Twelfth Night*: This is Elysium, Lady. There, such a metamorphosis is possible. The pagan paradise is introduced partly because the raptures of love are, so to speak, divine, and partly because only by casting erotic experience into such mythic terms can Carew purge it of numerous dissident elements, leaving, in a double sense, the pure sex of rapture. As the lover urges in *A Rapture*:

> Come then, and mounted on the wings of love
> Wee'le cut the flitting ayre, and sore above
> The Monsters head, and in the noblest seates
> Of those blest shades, quench, and renew our heates.
> There, shall the Queene of Love, and Innocence,
> Beautie and Nature, banish all offence

From our close Ivy twines, there I'le behold
Thy bared snow, and thy unbraded gold.
There, my enfranchiz'd hand, on every side
Shall o're thy naked polish'd Ivory slide.
No curtaine there, though of transparent lawne,
Shall be before they virgin-treasure drawne;
But the rich Mine, to the enquiring eye
Expos'd, shall ready still for mintage lye,
And we will coyne young *Cupids*.   (21-35)

The poem is genuinely erotic, and the eroticism is genuinely poetic. The latter condition derives in principle from Carew's poetic powers, and in practice from the very special conditions of "those blest shades" (*umbrae beatae?*) of Elysium. Sexual enjoyment becomes the characterization of a perfected life reigned over by Venus and purified by having "Innocence, / Beautie and Nature" banish "all offence," so sanctioning sexual delight.

Some may still think such matters too erotic to deserve discussion. But Carew exerts poetic power, an imagination for details, a tact in introducing them, and a mind of some capaciousness. Nature is there with Venus—an old association that Chaucer would have understood—but so, too, Innocence. The perpetual innocence implies a perpetual virginity welcoming perpetual ravishment; but Nature involves procreation, at least of "young *Cupids*." (Some interest lies in that "young" —what else?) Such paradox makes A *Rapture* the most genuinely poetic of all the erotic poems of the century: immortal innocence and immortal rapture.

My Rudder, with thy bold hand, like a tryde,
And skilfull Pilot, thou shalt steere, and guide
My Bark into Loves channell, where it shall
Dance, as the bounding waves doe rise or fall:
Then shall thy circling armes, embrace and clip

81

My willing bodie, and thy balmie lip
Bathe me in juyce of kisses, whose perfume
Like a religious incense shall consume,
And send up holy vapours, to those powres
That blesse our loves, and crowne our sportfull houres,
That with such Halcion calmenesse, fix our soules
In steadfast peace, as no affright controules. (87-98)

The erotic vision, both in itself and in its contribution to the
good life, ends "In steadfast peace." Donne sometimes contrasts the lovers' peace to the dissension outside their little
world.[38] Carew does not say whether that "steadfast peace"
is sexual, post-sexual, or unsexual in this passage, nor does
Lord Herbert of Cherbury in his *Ode Upon a Question
Mov'd*, nor Donne in *The Extasie*. But I should think that
with Carew it is the second. And yet, admitting that peace
to be post-sexual in the very limited sense of the coital, it
proves to be the finest "enjoyment" of all, the finest gift that
rapture can confer. And in the next chapter we shall have
ample reason to see that the steadfastness of any state, and
especially of one of peace, was that which the Cavaliers
sought most eagerly and that which they came to believe
would prove hardest to claim.

The next chapter (and the fifth) will also touch on the
pastoralism of such poems, but at this moment we must recall that James I took pride in his policy of peace. It may seem
to strain lyric poetry to jump as it were from "a bed / Of
Roses, and fresh myrtles" (*A Rapture*, 35-36) to the alarms
and excursions of the time. But such details fit into one larger
picture. On that canvas one can draw a scale of related experiences postulated on peace. To simplify enormously, we
may begin with *A Rapture*, go on to Lovelace's Lucasta and
Althea poems (themselves handsomely shaded), and pass

---

[38] See *The Extasie*, "The Canonization," "The Sunne Rising," and
"The Good-morrow."

without strain to the world of masques celebrating lords and ladies whose loves concur to affirm James's policy of peace. (Similar ranges could be drawn for experiences postulating a relation between love and war, as we shall later have reason to see.) But if my belief that peace is an important value for Carew is true, we should find it expressed in other contexts. Such a context will be found in a poem too little known. Aurelian Townshend had sent Carew an epistle on the death of Gustavus Adolphus (which created some poetic stir), to which he replied *In answer of an Elegiacall Letter upon the Death of the King of Sweden*. The news affects me, he says, but "Let us to supreame providence commit / The fate of Monarchs" (35-36)—sound Royalism but curiously negative expression. The "let" clauses build up to a splendid climax.

> Then let the Germans feare if *Caesar* shall,
> Or the United Princes, rise, and fall,
> But let us that in myrtle bowers sit
> Under secure shades, use the benefit
> Of peace and plenty, which the blessed hand
> Of our good King gives this obdurate Land,
> Let us of Revels sing, and let thy breath
> (Which fill'd Fames trumpet with *Gustavus* death,
> Blowing his name to heaven) gently inspire
> Thy past'rall pipe, till all our swaines admire
> Thy song and subject, whilst they both comprise
> The beauties of the *SHEPHERDS PARADISE*.
>
> (43-54)

There is, after all, no great distance to travel between "a bed / Of Roses, and fresh myrtles" where rapture leads to peace and a people "that in myrtle bowers sit / Under secure shades . . . / Of peace and plenty." Now it is not Venus who reigns, but "our good King," and life is improved by pastoral "Revels." Like the vision of love yielding in peace, so the vision of a land "Of peace and plenty" is a vision of a happy life.

iii. The Good Life

Such "enjoyments" as those Carew envisions show that the good life is good ethically and hedonistically. Enough of a Puritan lurks in us all to make us feel that ethics and hedonism, the *vita bona* and *vita beata*, do not exactly coincide. But we may look into the classical philosophers (not to mention our own hearts) and see that human happiness is by no means an eccentric ideal. Perhaps we must after all conclude that the Cavalier good life held in it ideals that have no necessary relation but which the poets related; in other words, that the ideal was conceived of as the eudaemonism of moral men. As Walton said, "an excellent Angler, now with God." We must understand that the moral epistles are founded on that happiness that can be derived by association with good people, friends. And we must also understand that Cavalier seduction poems and drinking songs have held appeal for three centuries, not merely because sex and drink still give pleasure, but because such poems raise their formal topics into visions of happiness. The world we look upon through Cavalier poetry is that of an England at peace (or hopefully at peace), dedicated to ancient rights of king and subject, liberal to friends and dependents, given to love, drink, song, angling and hunting, certain of the value of learning, and espoused (with certain infidelities) to the Anglican *via media*. We must look for ourselves to observe this very English landscape. Cavalier poems do not exclaim or point to themselves as do poems by the Metaphysicals—or by Milton. Nor do they seem to detain us by the toga, or lace, as do Dryden's. One stylistic feature of such good manners is the naturalness, the smoothness of verse.

Such qualities are genuine, and they are necessary characteristics of a poetry social in character. Only in recent generations has it been thought that art was related to advertising. The Cavaliers felt rather that the highest art concealed art. We see features of this seeming negligence or sprezzatura and of this hidden concern with art in the few surviving authors' manuscripts as well as in Jonson's account of what he lost.[39] And we see others in what "natural, easy Suckling" had to say, before "A Sessions of the Poets," of the faults of Carew.

> *Tom Carew* was next, but he had a fault
> That would not well stand with a Laureat;
> His muse was Hyde-bound, and th'issue of's brain
> Was seldome brought forth but with trouble and pain.
>
> (37-40)

Here are a couple of samples of Carew's laborious art, familiar to all. First, from the song, "Mediocritie in love rejected."

> Give me more love, or more disdaine;
> The Torrid, or the frozen Zone,
> Bring equall ease unto my paine;
> The temperate affords me none:
> Either extreame, of love, or hate,
> Is sweeter than a calme estate.[40]

Or, the first two stanzas of another song.

> Aske me no more where *Jove* bestowes,
> When *June* is past, the fading rose:
> For in your beauties orient deepe,
> These flowers as in their causes, sleepe.

[39] See A.N.L. Munby, "Phillips Back in the Sale Room," *The Times Literary Supplement*, 9 September 1965, p. 784, for a suggestion, so far unconfirmed, that a revised Herrick manuscript has been found.
[40] Lines 1-6; italics altered to roman letters.

> Aske me no more whether doth stray,
> The golden Atomes of the day:
> For in pure love heaven did prepare
> Those powders to inrich your hair.[41]

Such ease comes naturally only in the sense that there are some poets who know how to work for it. It implies that relations between people are so important that everything possible must be done to make communing possible.

The ease of the social mode varies considerably from Jonson's energy, to Carew's polish, to Suckling's negligence. Whatever its immediate stylistic atavar, such ease belongs to a homogenous society whose values are consonant from one aspect of the lives of the Cavalier poets to another. And such values centering on the good life lasted over a period from the end of the Tudor era well into the Restoration. We must assume that Cavalier experience was molded in the last days of Elizabeth and under James I, Charles I, and during the Interregnum. We may presume that manuscripts and song-versions circulated widely all that while. Yet it is a curious thing that the publication of books of Cavalier non-dramatic poetry is for the most part an Interregnum and Restoration phenomenon. Jonson's poems of course appeared earlier, but surprisingly late and seldom (1616, 1640, 1692). To take other poets, some of whom are but partly in the Cavalier camp, we must mention: Randolph (1639, 1640), Denham (1642, 1653, 1668), Waller (1645, 1664, 1690, etc.), Cowley (1646, 1678), Vaughan (1646, 1678), Herrick (1648), Lovelace (1649, 1657), Cartwright (1651), Stanley (1651), Benlowes (1652), Newcastle (1653, 1656), Hammond (1655), Henry King (1657, 1664), Sir Robert Howard (1660), Marvell (1681), and Cotton (1681, 1689). There is a certain paradox in the fact that Cavalier poetry is a creation of a period from about

---

[41] Lines 1-8; italic and roman usage reversed.

1595 to 1640 (not that it was not also written later) and that it is the possession of the period from about 1640 to 1680 (not that it was not circulating earlier). The paradox is a reminder that Cavalier poetry spans about a century from Jonson to Cotton's *Poems on Several Occasions*, and that its popularity during the Interregnum and the Restoration was owed to its protean social character.

The Cavaliers shared with Elizabethan and Restoration courtiers elements that have been called Royalism and Christianity, and something variously termed "humanism," "classicism," and "neoclassicism." For purposes of considering the good life, all three elements command attention. I shall try to discriminate but one, however, which I shall term classicism, if only to avoid associations entailed by the two alternatives. It cannot be said often enough that the nature of classicism depends entirely on the specimens offered or the models followed. Plato, Aristotle, Epicurus, and Diogenes offer four distinct philosophies. On the other hand, a Roger Ascham practicing a so-called Ciceronian prose style is no more classical than a Francis Bacon practicing the so-called Senecan. Donne's satires owe more to the Roman satirists than do Dryden's. Homer, Virgil, and Lucan differ markedly, and classical lyricism means very different things, depending upon whether the model be Pindar, Sappho, Catullus, or Horace. We must constantly remind ourselves that if such relatively small party-labels as "Elizabethan," "Metaphysical," or "Cavalier" must be regarded as capacious (or, as some would say, capricious) categories, then the geography of "Renaissance," and above all, of "Classical" is vast (and vague) indeed. So that the question posed for us by the classicism of the Cavaliers, and what may be taken to distinguish them from the Elizabethan, the Metaphysical, and Restoration poets comes down to this: on which classical writers was *their* classicism founded?

The classical basis of the good life may be represented most simply in terms of four poets: Martial, Anacreon, Claudian and, yes, Horace. To simplify yet further, Martial commonly provided the epigrammatic or moral bite needed to exclude what the Cavaliers disliked; he enabled them to brand perversions of their values, to say "All is not sound" when saying so was required. At the other end, Anacreon and Claudian admitted the pleasures of the small, the delights of wine and love, and the repose of a natural or country life.[42] Anacreon's and Lovelace's and other poets' grasshoppers might freeze in a green blur, but something was learnt all the same. Just as Martial at his highest and Claudian in his *Old Man of Verona* contributed to the classical resonance of poems on the country and estates, so Anacreon plain and Anacreon applied had more philosophical import than Anacreontic verse at first suggests. All this being said, it remains true that the classical poet central to Cavalier experience is Quintus Horatius Flaccus.

It may well appear to some, as it long did to me, that neo-Stoicism defined for the Cavaliers the *vita bona* and neo-Epicureanism the *vita beata*. There is something in the hypothesis, but it presumes the importance of Seneca, Lipsius, and other worthies, when in fact from 1580 to 1630 there is a falling-off of interests in those writers commonly identified with neo-Stoicism in England.[43] An individual can of course

[42] The true significance of Martial, Anacreon, and Claudian to Cavalier poetry requires further study, concern with transmission of editions, translations, echoes, and use of formulas and motifs. Claudian is probably least familiar in this discussion, but his minor poems were important as well as *The Rape of Proserpine*. Even the panegyrics had influence. But my sketchy remarks derive from simply reading through the poets, classical and Cavalier, and their editors, not from proper or fully informed study. The next chapter will return to poetic features of Cavalier "classicism."

[43] See "Patterns of Stoicism in Thought and Prose Styles, 1530-1700," *Publications of the Modern Language Association*, LXXXV (October,

be interested in Buddhism when most of his religious contemporaries are Christians and Jews. Jonson's images of the pitcher of the soul and of the turbulent waves can, for example, be found in neo-Stoic writings on constancy.[44] But what if such imagery is found also in Horace, and what if Horace is repeatedly published during our period and Lipsius is not? Since both these conditions obtain, it is Horace that must concern us.

There is no question but that Horace was concerned with the good life (*vita bona* and *vita beata* alike) throughout most of his work. But how did he define the good life philosophically? Classicists have been debating the matter for centuries. Seventeenth-century writers as late as Dryden debated it; so do scholars today. Some have thought Horace an Epicurean, others an Epicurean converted to Stoicism by the famous thunderclap in a clear sky (*Odes*, III. v), and others have regarded him as a deliberate Eclectic. The problem is of some importance to us, because the philosophical basis of Cavalier poetry is to a considerable extent that of a Christianized Horace, or of a Horatianized Christianity. The question is therefore not one, but two: we may debate whether Horace himself adhered to this or that philosophy; and we must ask how he was read by commentators and poets in the seventeenth century. I suspect that the best answer to the debate (from what one knows of poets and of other philosophically inclined Romans of his time) is that Horace was an

---

1970), 1023-1034. As I say in that article, my doubts about usual accounts of tides of Stoic influence and details of the classical figures involved began when I compared the two editions of the first volume of Maren-Sofie Røstvig's excellent study, *The Happy Man* (Oslo and Oxford, 1954, 1962). What disagreements may separate Professor Røstvig and myself do not affect my opinion that her study is the most useful survey we possess of a seventeenth-century poetic tradition apart from the Metaphysicals, and it certainly bears centrally on this chapter.

[44] See *ibid.*, 1st ed., I, 50-51.

Eclectic. And certainly the Horace of the commentators is eclectic in another sense of being interpreted by Christian writers anxious to find as few conflicts as possible between his ideas and their own, and probably anxious, too, to bring to bear on Horace that complex of Christianity in classic guise —with elements of Stoicism and neo-Platonism absorbed beyond distinction. We know that Horace was often treated in Stoic terms,[45] and that his Sabine farm spoke home to the Cavaliers, for "the happy country life [was] the most typical expression of the Royalist and Anglican spirit of the seventeenth century."[46]

The quality that above all distinguishes the "Stoic" element in Horace and the Cavaliers is a self-sufficiency involving self-knowledge and wisdom, living a life relying on oneself, and refusing to be distracted by the baubles of the world. The self-sufficiency is part of the happy country life: *Beatus ille* (*Epodes*, ii, "Happy the man . . .") . Such a man, as Jonson and Selden remind us, prizes the few and detests the most: *Odi profanum vulgus et arceo* (*Odes*, iii. i, "I hate the uninitiate crowd and keep them far away") . One must fly, as Jonson recommends his friend, from the immoral pomp of the world: *fuge magna* (*Epistles*, i. x, 32, "Flee grandeur . . .") . Indeed the whole world is mad except for the calm wise man: *Satires*, ii. iii.[47] The question is, *Quisnam igitur*

---

[45] See the plates in *ibid.*, 2nd ed., i, ch. i.

[46] *Ibid.*, 1st ed., i, 22. The phrase, "the Royalist and Anglican spirit," is another way of talking about Cavaliers, at least until certain Restoration alterations.

[47] This satire is Horace's longest and perhaps best sustained. As the commentators show, it is based on the Stoic aphorism that all are mad except the wise man: πᾶς ἄφρων μάινεται; or, as the Delphin edition put it in its label, "Fuse tractat paradoxum illud Stoicum: Omnes stultos insanire." (*Quinti Horatii Flacci Opera*, "Editio Decima" [London, 1790], p. 446.) The same heading will be found in Minellius's edition (my copy, Rotterdam, 1677), p. 352; and other editions use similar expressions. My translations of classical authors in this chapter are taken (unless other credit is given) from the very useful Loeb Classics editions.

liber? The answer is, *sapiens, sibi qui imperiosus (Satires,* II. vii. 83. "Who then is free? The wise man, who is lord over himself). Once again the thought is based on Stoic aphorism that only the sage is free (ὅτι μόνος ὁ σοφὸς ἐλεύθερος). Or, to take the doctrine of self-sufficiency as it is sustained best in Horace, one may look at the famous *Nil admirari* epistle that Pope was to adapt (*Ep.* I. vi, "Marvel at nothing . . ."). There we learn that only self-sufficiency offers assurance of happiness and virtue. And it is altogether typical of Horace that commentators should emphasize that behind this important poem there should lie not only Strabo's conception of philosophers in general and a remark by Pythagoras, but also the imperturbability (ἀθαμβία) of Democritus, the calm (ἀταραξία) of the Epicureans, and the freedom from emotion (ἀπάθεια) of the Stoics.

All of us recognize these strains as central ones in Cavalier poetry, and indeed the question of the good life at its most crucial is not so much what it is as how to keep it. Repeatedly, the world outside is in storm, is very cold, or is in flood (images found together in Horace, *Odes,* I. ii), or else the pressure of the world requires flight to one's integrity or to the country (which often is very nearly the same thing). Not only self-sufficiency but sanity in a mad world and freedom among people self-enslaved are required. And in the second half of the century, the question becomes one of flight to the country; *fuge magna,* indeed, but with greater sense of urgency. From Cowley, from Bunyan, from Dryden, and from Restoration drama we hear the question, "Whither shall I fly?" Writing toward mid-century, Dudley North asks the same question in *A Forest of Varieties* (1645).

> With whom shall I converse [i.e., spend my time]? Where is the dwelling of wisedome, or innocent simplicitie of heart? . . . Justice [i.e., Astraea] long since tooke her flight to heaven, Peace, wisedome and integritie have followed

. . . I will shelter my selfe as well as I may: naturally I love stirring, but the weather must bee fairer.[48]

From Jonson to Cotton, changes are rung on such ideas. The recurrent images of shelter and warmth vs. cold and storm, of making love or drinking with friends, of turning in on oneself and away from the world in order to be free: such and similar formulations are perhaps less philosophical ideas, or even less motifs, than rites and icons.

The numerous editions of Horace published between 1580 and 1640 (and indeed before and after, although that period was the peak) show that he was thoroughly read, and so it is not necessary to search into learned crevices or to construct elaborate hypotheses. The appeal of Horace to Cavalier poets could not have been lessened by virtue of the fact that he was himself a poet and a courtier. On the other hand, we owe it to our writers to grant that they read more widely than in Horace—Stanley after all produced a massive history of philosophy. And I shall therefore endeavor to deal with a much confused subject. In the seventeenth century the two most popular of classical writers who may in some way be classed as philosophers were Seneca and, above all, Cicero.[49] Seneca's *Works* "morall and naturall" were published in 1614 and 1620. Among Seneca's moral writings, the one that naturally

[48] Quoted by Røstvig, *The Happy Man*, 1st ed., I, 54.

[49] Richard Ashcraft has surveyed the contents of a number of book collections made during the seventeenth century and sold at auction late in that century or early in the next: by a wide margin, Cicero is the dominant classical philosopher owned by representative seventeenth-century authors. Ashcraft, a colleague in the Department of Political Science at UCLA, has kindly given me information from as yet unpublished materials. But much of what he will ultimately publish lies behind his very interesting article, "John Locke's Library: Portrait of an Intellectual," *Transactions of the Cambridge Bibliographical Society*, v (1969), 47-60. One thing emerging strongly from his work is that the pattern that I have observed (see n. 43, above) from study of the short-title catalogues holds even when imported books are taken into account.

fits in with the purposes of this chapter is *De Vita Beata*, which, however, is closer to the *vita bona* as I have defined it. Integrity is required, and so is attention to the advantages of life (Seneca was rich), but without too much love for them. This discussion (III. 3-4) contains the phrase *sine admiratione* (for the good things) which editors naturally compare with the Horatian *Nil admirari*. In a long passage (XII. 3-XIII. 3), Seneca agrees with Epicurus as he essentially was (in Seneca's view)—an advocate of what is reasonable, of study and following of nature, of content with little—though he disagrees with the voluptuaries who term themselves Epicureans. Since in the *De Otio* he goes so much farther toward Epicurean doctrine as to advocate, for the Stoic, retirement from public life, one can see that one will find no pure Stoicism in Seneca. Indeed, Seneca was far more popular for his so-called chorus on moderation in the *Thyestes*, Act II.[50] And one would be hard pressed to show that that passage was more Stoic than is the poetry of Horace. It is Cicero, after all, that must concern us.

For reasons that are not very clear, students of English literature have set up an opposition between Seneca the Stoic with a responsible prose style and florid Cicero with no philosophy at all. Leaving aside the question whether Seneca's style is not, as the classicists say, Asian, Cicero's philosophy is centrally that of the Academy, developing out of Plato and Aristotle into a basically skeptical tradition in matters epistemological, doubting that more than probablism was open to man, although on ethics and theology he affirmed the Stoic position. On the other hand, he was less a philosophical innovator than perhaps the most important transmitter of such major classical philosophical traditions as Platonism.[51] And

---

[50] See Røstvig, *The Happy Man*, 2nd ed., 1, 21 ff. *Thyestes* was published five times between 1580 and 1640.
[51] See Pierre Courcelle, *Late Latin Writers and Their Greek Sources*

later in his life he wrote a number of works that are highly Stoic in emphasis. It will be useful to glance at a few of these, considering them as works varying in their Stoicism and varying in their popularity between 1580 and 1640 (which I take to be the formative period of Cavalier ideas). That lengthy Stoic work, *The Tusculan Disputations,* was published just twice: as often as the *Works* "morall and naturall" of Seneca. During the period, *De Senectute* was published seven times, as often as the *Works* and *Thyestes* of Seneca together. This treatise on old age was popular with Christians partly for its propounding of the immortality of the soul—using as authorities Pythagoras and Plato, whose philosophy is summarized in brief (xxi. 77-78). There is a long section (xv. 51-xvii. 59) on the joys of country life that must have echoed in the breast of Cavalier poets. Cicero's speaker in the dialogue, Cato the Elder, speaks lovingly of the country, at first chiefly in a Georgic vein. "I might enlarge upon all the many charms of country life" (*permulta oblectamenta rerum rusticarum* [xvi. 55]), he says, and indeed he does go on to do so.

> For my part, at least, I am inclined to think that no life can be happier [*beatior*] . . . not merely from the standpoint of the duty [*officio*] performed, which benefits the entire human race, but also because of its [that of the life of the farmer] charm already mentioned, and the plenty and abundance it gives of everything that tends to the nurture of man and even to the worship of the gods. (xvi. 56)

As we shall see, the concept of "offices" is an important Stoic idea. But there are one or two other passages in *De Senectute* that make it a work central to the tradition of "the happy man" in the seventeenth century. Cato exclaims, at one point (xiv. 49), for example,

(Cambridge, Mass., 1969), pp. 66 and 165 ff. St. Augustine's rhapsody on reading Cicero's *Hortensius* is of course a commonplace.

But how blessed it is for the soul, after having, as it were, finished its campaigns of lust and ambition, of strife and enmity and of all the passions, to return within itself, and, as the saying is, "to live apart"!

Here, in little and in prose, we find the central principle of Jonson's poem to Wroth.

> How blest art thou, canst love the countrey, *Wroth*,
>   Whether by choice, or fate, or both;
> And, though so neere the citie, and the court,
>   Art tane with neithers vice, nor sport. (1-4)

After a review of the delights and plenitude of the country and a contrasting scene of the evils of city and court, Jonson turns to Sir Robert living happily to himself (*sibi vivere*) in the country, an ideal as much Epicurean as Stoic.

Another popular work by Cicero, *De Amicitia* (published seven times in our half-century), touches on matters that have often been treated in studies of seventeenth-century poetry but in terms of their having been derived from writers far less familiar. Laelius, Cicero's major speaker in this dialogue "On Friendship," very early (II. 8-10) urges the Stoic doctrine of equanimity, of self-possession and control of one's passions. And one of the central virtues derived from consideration of friendship is constancy. It is introduced in order to speak of the loyalty necessary between friends (XVIII. 65). But such virtues become the basis for distinguishing between the *"Most"* that Selden and Jonson spoke of and the *"Few."* Cicero contrasts the giddy citizens with "one who has stability, sincerity, and weight." The Latin phrase (xxv. 96) is *constantem et verum et gravem*. And the concluding section of the dialogue begins with a discussion of virtue: "For in Virtue is complete harmony [*convenientia rerum*], in her is permanence, in her is fidelity [*constantia*]" (XXVII. 100). Not

95

only is all this standard Roman Stoicism, but also the ethical basis of Jonson's poetry and that of those who follow him in writing of friendship. It may well be true that many poets of the time were reading and becoming excited by Lipsius on constancy. The *De Constantia* was published twice in Great Britain during our half-century, once in Latin (1586) and once in English (1595). Perhaps Lipsius was important. It is certainly true that Cicero's enduringly popular essay on friendship, which (as the last chapter of this book shows) is a major Cavalier theme, was published seven times in large editions.

The last work by Cicero that we must glance at is the most obviously Stoic of his writings, *De Officiis*, "Tully's *Offices*," as the work was often referred to, concerns the Stoic concept of duties or obligations. This one work was published as often (twelve times) as the total canon of Seneca, and since it was included (with *De Senectute* and *De Amicitia*) in the Stationers' Patent for school texts, the number of copies printed was probably several times in excess of the usual printing.[52] *De Officiis* is one of the truly popular works of our period, and it illustrates better than any other evidence the degree to which Cicero was responsible for communicating Stoic philosophy (as indeed other kinds) of the seventeenth century. The lengthy discussion of the Stoic "offices"

---

[52] I am indebted to Mr. Arthur Crook, editor of *The Times Literary Supplement*, for putting me in touch with an anonymous reviewer who has pointed out that the works by Cicero cited in Pollard and Redgrave's *Short-Title Catalogue*, nos. 5266-5273 were part of a School Book Patent. From 1635, *as far as is known*, such Patent books would be printed in impressions of 5,000 or 6,000 copies, as opposed to the normal 1,250 or 1,500. From 1587 to 1635 the multiple was *probably* smaller. For the complex evidence, see W. W. Greg, *A Companion to Arber* (Oxford, 1967), pp. 95, 266-68; and Cyprian Blagden, *The Stationers' Company* (Cambridge, Mass., 1960), pp. 186-87. Professor D. F. McKenzie of the Victoria University of Wellington has put me in his debt with a very informative letter that causes me to italicize certain words in preceding sentences.

is based on the Περὶ τοῦ Καθήκοντος of the Stoic, Panaetius of Rhodes. The main topic is introduced most evidently at I. 4, but discussion of the moral duties runs through the whole work. I cannot see a need for explicating this work or even its relevance to seventeenth-century thought, when the one task would be so disproportionate and the other so unnecessary, given the obvious relation. But one may perhaps dip his hand into the stream a time or two. How like Jonson Cicero may sound! After agreeing with the Stoics that courage is a virtue that champions the right (I. xix. 62), and with Plato that knowledge must be joined with justice, he concludes:

> And so we demand that men who are courageous and high-souled [*magnanimos*] shall at the same time be good and straightforward, lovers of truth, and foes to deception; for these qualities are the centre and soul of justice. (I. xix. 63)

In Jonson's more concentrated version, "For he, that once is good, is ever great."

The strong ethical tone will make clear the difference between *De Officiis* and the other Ciceronian works we have been glancing at. *De Senectute* rather surprisingly treats of the happy country life, and *De Amicitia* mingles concern with the happy and the good lives; but *De Officiis* concerns the good life and its requisites. In the last, Cicero sets out the Stoic ideals in a crucial series of sections (I. xx-xxi and xxvi). It is in the last book, however, that he speaks explicitly about a matter of direct concern to this chapter, the good man, the *vir bonus* (III. xix). There is nothing startling about his description: "hee is a good man who doth good to whom he may: and hurteth no bodie, but provoked by injury."[53] The good man does what is right when he can.

The Cavaliers felt themselves provoked often enough as

---

[53] *Marcus Tullius Ciceroes three books of duties* (n.p., 1583), fol. 145ᵛ. Or, more clearly, Roger L'Estrange's translation: "only He is a *Good* man, who does as much good to *Others* as he can, and harms *no*

individuals in our sorry world; and they felt injury done them during the 1640's and 1650's. But rather than take the famous Stoic measure, suicide, they took to refuge in friends, in the country, or in themselves. As we have seen, that may not be pure Stoicism, but it agrees with Cicero's Stoic works. And it agrees with Horace as he expresses himself in poem after poem, not least in his magnificent ode, "Tyrrhena regum progenies" (III. xxix).

> Happy the Man, and happy he alone,
> He, who can call to day his own:
> He, who secure within, can say
> To morrow do thy worst, for I have liv'd to day.
>
> (trans. Dryden, 65-68)

Or again,

> Content with poverty, my Soul I arm;
> And Vertue, tho' in rags, will keep me warm.
>
> (86-87)

Of classical writers, Horace and Cicero best exemplify the shadings of meaning attached to the good life. As we shall have occasion to see in the next chapter, the poetic approaches possible also owed much to Horace. And the philosophical basis is most conveniently described as a Ciceronian eclecticism. He had studied the major philosophic schools of his time—the Academic, Epicurean, and Stoic—and his brother Quintus favored the Peripatetic. It cannot be said often enough that Cicero is a transmitter of ideas, for if he innovated very little, what he passed on was a sense of what his own experience had shown him to be important to life. There is a fundamental sense that, practical Roman that he was, Cicero sought out not the disinterested abstract wisdom

---

body without some Injurious provocation" (*Tully's Offices* [London, 1680], p. 181).

of the Greeks, but knowledge of the good life. So, too, the Cavaliers. Significantly, his major effort to do so came when he had to retire from political activity, when an unfavorable regime led him to retreat to the country, to his own moral and other resources. Again, so, too, the Cavaliers—after Jonson had shown the way. Cicero's philosophical writings appeared in rapid succession during a period when he was sick at heart over the downfall of the old political order he had done so much to uphold, when he was troubled by debt, and when he was afflicted by such domestic sorrows as the death of his beloved daughter, Tullia. On 7 December in 43 B.C. he was, in modern parlance, purged—ordered killed by Octavian to please Mark Antony. These familiar matters may seem remote from Cicero's warm philosophical writings and from Cavalier poetry. But both concern the good life and, as we must now observe, the Cavaliers as well as Cicero had good reason to think a life combining pleasure and morality was in danger of being lost. In fact, all that has been said of Cicero in these paragraphs can be applied, *mutatis mutandis*, to the Cavaliers. And in such application lies both the truth of recurrent human experience and a lasting significance of the classical past. In their concern with the good life, the Cavalier poets treated an ideal of central and enduring human importance.

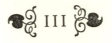

THE RUINS AND REMEDIES
OF TIME

*O you immortall gods, what is there in mans
life, that is of any long continuance?*
> —Cicero, *De Senectute*, trans. 1648

*All things in this World are so frail and
uncertain.*
> —Cicero, *De Amicitia*, trans. 1691

He lives, who lives to virtue.
> —Herrick

IF WE THINK that the good life envisioned by the Cavaliers
was something too good to last, then we have partaken of
their vision, or have looked about this world a bit. One of the
differences between our experience of time and transience and
that of the Cavaliers is that we do not share the emblems they
shared with their contemporaries, their ancestors, and antiquity. Flowers blossom and quickly fade. *Sic vita.* The
grasshopper sings through summer and freezes in the cold.
The child is born and the man soon dies. We do have a sense
of these rhythms, but rather than employ them as commonplaces we seek to embody them in such more abstract explanations as the sociological. We might, for example, choose to
say that the Cavalier urge to seize the day, or the Cavalier
regret that we are but decaying, stemmed from an awareness
that the class they represented was living on the margin and
about to go under, as "Puritanism" or "the rise of capitalism"
came to bury the old order. But I find no evidence for this.

Between Cicero and Cotton, the sense of the ruins of time is continuous, and the experience of the seventeenth century was not uniquely unsettling. When had England been settled and calm? It had not been long since that the Wars of the Roses had ended, that the Reformation had divided England, that Henry VIII was succeeded by queens of doubtful legitimacy, and that the threat of invasion from Spain had been so frightening that, according to the old story, Thomas Hobbes's mother miscarried and brought him prematurely into the world. The peace policy of James I may have been vain, and the efforts by Charles I to govern without Parliament were certainly wrong. But men's illusions are not held the less passionately for their being false.

There may be one respect in which the Cavaliers were nostalgic and retrospective, unlike some of their contemporaries. That is, their social order was one requiring a continuing of their ancestry of titles, their concept of social legitimacy, and their rights. Conservativism was instinctive. Their poetry reflects such an assumption, as Metaphysical poetry with its attention to the now and the ever of life does not. Also the increasing millenarianism of Puritanism during the early half of the century certainly led such "new men" to look ahead rather than back. And yet Andrew Marvell, who is as surely some kind of Puritan as he is some kind of Cavalier, heard Time's winged chariot behind him, and had less of a sense that the millennium and the conversion of the Jews were at hand, than he did of the deserts of vast eternity lying before. And what Marvell shared with Herrick, they both shared with Spenser and Shakespeare and with generations of writers. The frailty of the human estate, or man's subjection to time, are commonplaces of literature because they are commonplaces of life. And it requires no unusual powers of human psychology to understand that the fear of time might motivate man to "enjoyments" of the happy life; just as that to men whose ideals were finally Christian, the good life gave promise that the good man could defeat time with eternity.

i. Time and the Good Life

The sense of time's pressure was no new thing in the seventeenth century. In our tradition, the pulse of time begins to beat in the poetry of the Greeks, and in the Asian traditions the counterparts have their own store of commonplaces. Any talk about the role of time in Cavalier poetry must, therefore, consider what is distinctive about the Cavalier treatment of time. The distinction resides partly in the prominence of this theme among others, partly in its connection with earlier native and classical traditions, and partly in its function as a feature distinguishing Cavalier poetry from Metaphysical. To take the last first, it is well known that if Donne is anti-Petrarchan, he still knew the Petrarchan conventions well enough to move among them with ease. Floods of tears, tempests of sighs, blazons of a lady's features, complaints, valedictions, palinodes—all these decorate his verse, often in extraordinary revisions. Similarly, Donne uses that form which the Cavaliers, like him, inherited from the Elizabethans: the seduction poem. Donne's poem, "The Baite," is but one of many parodies of Marlowe's "Come live with me, and be my love." (Walton puts both into *The Compleat Angler*, Donne's perhaps because of its piscatory imagery.) "The Flea" is certainly, wittily, and outrageously a seduction poem, and the second half of *The Extasie* is thought to be such by some readers. And yet all these poems fail to employ the prime argument used by the Cavaliers.

> Come my *Celia*, let us prove,
> While we may, the sports of love;
> Time will not be ours, for ever:
> He, at length, our good will sever.

Spend not then his guifts in vaine.
Sunnes, that set, may rise againe:
But if once we loose this light,
'Tis, with us, perpetuall night.[1]

Like Campion before him and Alexander Brome after him, Jonson levies on Catullus, v. 1-6. Whereas in Donne's "lovers seasons" the sun is a busy old fool who may be ordered about, to Catullus, to Campion, and to Jonson, the problem is rather that the sun continues to shine, and the rain to fall, on the just and the unjust, but not on the dead. The hard season of winter blows in; the dark time of night knows no end. *Nox est perpetua,* "if once we loose this light," this chance, this hour. Donne and his followers move outside the main humanist, Renaissance line in their use and sense of time, whereas Jonson and the Cavaliers move very much in it. We may recall Shakespeare's "golden lads and girls" and their inevitable ending. We remember how the same great poet uses the argument of natural analogy: "That time of year thou mayst in me behold . . . ."

Nor is Donne eccentric among Metaphysical poets in his seeming indifference to time. We shall shortly observe how frequently flowers, and especially roses or daffodils, are emblems of the transience of beautiful things. In his poems like "Life" and *The Flower,* however, George Herbert manages to use the emblem with its *sic vita* overtones and yet without playing on the ruins of time.[2] Time and Death are certainly not absent from the world-view of the Metaphysical poets, but neither exerts its force in their poetry as an unconquerable enemy of man. Rather, one must fear the world, the flesh, and the devil, the first in the context of their secular poems,

[1] Jonson, "Song. To Celia," ll. 1-8.
[2] I have discussed that poem at some length in *The Metaphysical Mode from Donne to Cowley* (Princeton, 1969), ch. v. ii.

and all three in their religious. I think that readers will discover very often that poems they think on other grounds, or
by instinct, to be characteristic Cavalier poems will possess
the temporal theme, and that poems like *To his Coy Mistress*
that seem partly Cavalier are often so precisely in respect to
their awareness of time. The very world of the Cavalier good
life includes, if not precisely some *memento mori*, then a
memento temporis, and the reminder of time is often a suggestion of death.

> So when or you or I are made
> A fable, song, or fleeting shade;
> All love, all liking, all delight
> Lies drown'd with us in endless night.
> Then while time serves, and we are but decaying;
> Come, my *Corinna*, come, let's goe a Maying.[3]

So concludes one of those seemingly perfect Cavalier poems
with its "fleeting shade" and, in Jonson's phrase, its "vision
of delight." But there is also fear of "endless night," once
again that Catullan *nox . . . perpetua*, and a conviction that
"we are but decaying."

These few remarks will already have argued the prominence
of the temporal theme as well as its function as an element
distinguishing Cavalier from Metaphysical poetry (or rather
Metaphysical from other sixteenth- and seventeenth-century
styles). The rest of the chapter will continue to stress the
prominence of the theme, and we may now seek to understand something of such importance by considering its origins
and distinguishing its properties. The theme, if not the Cavalier tone, of the depredations of time takes its fullest classical
expression in the last book of the *Metamorphoses* (xv. 176

[3] Robert Herrick, *Corinna's going a Maying*, ll. 65-70.

ff.). Many of the important images are there: time as a flow, a sea, a river (xv. 176 ff.); time as a circle or cycle (xv. 184 ff. and *Metamorphoses, passim*); the emblematic sun and moon (xv. 186 ff.), the emblematic seasons of life (xv. 199 ff.); and the extraordinary changes in even those things that seem most constant (xv. 237 ff.). Above all, Time is a personified, named villain, the devourer of things—*Tempus edax rerum*—along with a second villain, envious Age—*invidiosa Vetustas* (xv. 234). Such phrases are so traditional, such commonplaces in the sixteenth and seventeenth centuries, that one would be hard-pressed to find a modern equivalent. Suffice it to say that they were so well known that when Charles Cotton chose "Tempus edax rerum" as a subtitle to a poem, "On Marriot" (*ca.* 1653), the poem turns out to be a mock elegy.

What distinguishes the Cavalier usage from that of other styles is the concern over the good life, both its acquisition and its loss. To the extent that Time pitted himself against the happy life, one might take such measures as seizing the day, making much of time, enjoying freely. And yet, especially in the love poems urging just such courses, the valiant effort to enjoy seems doomed, and the man who presses on his lady the argument of time for her to sport or enjoy often persuades himself of the reason without getting the lady to accept the inference. The accent in such poems varies greatly, and the response may differ. But the lesson admits no doubts. Herrick urges well:

> Gather ye Rose-buds while ye may,
>   Old Time is still a-flying:
> And this same flower that smiles today,
>   To morrow will be dying.

Or again:

Come, let us goe, while we are in our prime;
And take the harmelesse follie of the time.
  We shall grow old apace, and die
  Before we know our liberty.[4]

Carew advances similar arguments, but with a dialectic borrowed from Donne. His "Song. Perswasions to enjoy" possesses two hypotheses, a kind of Ramist either / or. One possibility is temporal: "If the quick spirits in your eye / Now languish, and anon must dye . . ." (1-2). The second possibility is pastoral: "Or, if that golden fleece must grow / For ever, free from aged snow . . ." (7-8). Either way, the conclusion is that the woman should "enjoy":

Thus, either *Time* his Sickle brings
In vaine, or else in vaine his wings. (13-14)

We can see from a much longer poem, "To A. L. Perswasions to Love," that it is the temporal theme rather than the dialectic that is consistent in Carew. Although too lengthy to quote from extensively, the urgency of the ending is too remarkable to omit.

Oh love me then, and now begin it,
Let us not loose this present minute:
For time and age will work that wrack
Which time or age shall ne're call backe.
The snake each yeare fresh skin resumes,
And Eagles change their aged plumes;
The faded Rose each spring, receives
A fresh red tincture on her leaves:
But if your beauties once decay,
You never know a second *May*.

---

[4] Herrick, "To the Virgins, to make much of Time," ll. 1-4; *Corinna's going a Maying*, ll. 57-60.

Oh, then be wise, and whilst your season
Affords you dayes for sport, doe reason;
Spend not in vaine your lives short houre,
But crop in time your beauties flower:
Which will away, and doth together
Both bud, and fade, both blow and wither.

(69-84)

Other tones will be found in Waller's beautiful reminder to the lady by way of address to a flower representing her, "Go lovely Rose!" And we have seen (in the first chapter) how, like Carew, Charles Cotton threatens with time an "ingrate-full beauty" in "To Chloris. Stanzes Irreguliers."

We can, then, find abundant evidence in the poems to show two things: the widespread idea that time was inimical to the good life, with the concomitant belief that by seizing advantage of the hour one might to some degree defeat time; and a considerable variety of tone in poems using such "per-swasions to enjoy." But what is essential to this theme, and how does it most meaningfully relate to the good life? The problem is one of very fine shading as well as of the presence of shade. Perhaps a knowledge of the widespread character of the temporal theme will enable us to seek out the Cavalier emphasis and enable us to settle firmly the effect of time on the good life.

There seems to me no need to consider philosophies of time, and there is not much help to be found from most of the Greeks.[5] The Roman love poets are the ones who set the

---

[5] The sense of transience and time's pressure will be found occasionally in the letters of "Philostratus" (e.g., no. 17 in Loeb ed., *Alciphron*, etc.), and special mention must be made of Anacreon, although his stance was, as we shall see, at variance from the Roman. Anacreon could be found in editions of the Planudean *Greek Anthology*; the Palatine manuscript discovered by Salmasius in 1606 was of course not published till the eighteenth century.

most usable example of such motifs as roses, transience, seizing the day, and so on.[6] But the motifs will be found as well
in the *Georgics* and the *Aeneid*, in Cicero and Seneca.[7] Montaigne echoed the concern in "Of Glory."[8] The theme is
embodied in such narratives as Shakespeare's *Venus and
Adonis* (129, 132) and Sydney's sonnet sequence, *Astrophil
and Stella* (4th song, st. 5). It is part of Renaissance epic, with
its greater concern with love: in Tasso, *Gerusalemme Liber-
ata*, xvi. 14-55, and following him, Spenser in *The Fairie
Queene*, ii. xii. 74-75. Burton went on, in his fashion, about
it in the *Anatomy of Melancholy* (e.g., iii. ii. 5. 5), and
Comus runs through the imperative, "be not coy" (737 ff.)
and the motif of transience (743-44). So cursory a sample re-
assures us on at least one head, namely, that the song of time
as a persuasion to enjoy is one of the good old songs. But we
can also see this difference: traditionally, the *positive* employ-
ment of such themes is the possession of lyric poetry. In other
forms, the status is ambiguous or doubtful. More than that,
most of the examples given show that love is one aspect of
the good life dealt with in lyric poetry making use of the
arguments of time. But it is not only love that is to be enjoyed, and by looking more closely at two classical poets, we
can observe how Cavalier poets were able to fashion actual
poetry out of the great commonplaces.

The two classical poets are Anacreon and Horace, both of
whom were mentioned for their ideas about the good life in
the last chapter. It is now the poetic construction of the good

[6] See Catullus, v. 4-6; lxii. 39-48; Ovid, *Ars Amatoria*, iii. 65-66;
Tibullus, i. viii. 47-48; and Propertius, iv. v. 39-60.

[7] Virgil, *Georgics*, iii. 284; *Aeneid*, vi. 275; Seneca, *Epistulae*, cviii.
24 ff.; *Hippolytus*, 446-51, 773-76; Cicero, *De Senectute*, xix. 69; *De
Amicitia*, xxvii. 102—these last two being translated in part at the head
of this chapter.

[8] In the translation by Charles Cotton, *Essays*, 3 vols. (London,
1686), ii, essay xvi.

life that concerns us. There had been earlier and would be numerous later recollections of Anacreon, but Thomas Stanley's translation of the Greek poet, "Printed in the year, 1651," is the most complete and most accurate.[9] Number xi in Stanley's series, "The old Lover," presents Anacreon in the role in which he was pictured by seventeenth-century poets.

> By the women I am told
> 'Lasse *Anacreon* thou grow'st old,
> Take thy glasse and look else, there
> Thou wilt see thy temples bare;
> Whether I be bald or no
> That I know not, this I know,
> Pleasures, as lesse time to try
> Old men have, they more should ply.

Cowley's version, "Age," ends somewhat differently.

> This I know without being told,
> 'Tis time to *Live* if I grow *Old.*
> 'Tis time short pleasures now to take,
> Of little *Life* the best to make,
> And manage *wisely* the *last stake.*

Both versions show that if time is the enemy of the happy life, the happy life possesses possibilities as a remedy of time. Those of time's remedies that are less desperate than another quick bowl of wine will concern us later; for the moment Anacreon (and Thomas Stanley) help me show that the Cavalier values raised the Cavalier problems, and that, to some extent, the problems contained their own solutions.

[9] See Stanley, *Poems*, 1651. The Clark Library copy includes *Anacreon. Bion. Moschus* [etc.] as a separately paginated work. The 55 poems appear on pp. 3-30, and Stanley's interesting commentary, on pp. 81-112. It is a pity that there was not space in Galbraith Miller Crump's valuable edition of Stanley (Oxford, 1962) to include Stanley's notes, but Crump shows how Stanley proceeded as a translator.

> I am sprung of humane seed,
> For a lives short race decree'd;
> Though I know the way I've gone,
> That which is to come's unknown;
> Busie thoughts do not disturb me;
> What have you to do to curb me?
> Come, some Wine and Musick give;
> Ere we dye, 'tis fit we live.

Here (*Anacreon*, no. xxiv) we see the Anacreontic vision: full awareness with a shrug of indifference. The first four lines could take almost any direction, given various seventeenth-century styles; and the possibilities raised before the last two lines are not confined to Anacreontic verse. One of the problems of defining seventeenth-century Stoicism or Epicureanism is that the self-sufficiency, *sibi vivere*, belongs as much to one Roman or English vision as the other. Stanley's translation of Anacreon might, then, have turned at its end to the sweet urgency or ethical imperatives of Herrick, to the negligence of Suckling, or to the somber vision of death in Marvell's *To his Coy Mistress*.

Perhaps, in the end, Anacreon's self-awareness proved most useful to our poets. It is remarkable that many of them, and often the Metaphysicals as well, presented another feature of the Anacreontic self-awareness, the tendency to deal with the small, a tendency that has been called an agoraphobia.[10] A number of Claudian's poems on small animals must have proved congenial, too, but it was Anacreon (no doubt more often in Latin translation than in Greek) who initiated the poetic fashion for treating small creatures: "The Dove" (ix), "The Swallow" (xii, xxxiii), "The Bee" (xl), and our old friend, "The Grassehopper" (xliii). (The titles are Stanley's.)

---

[10] See Toshihiko Kawasaki, "Donne's Microcosm," in *Seventeenth-Century Imagery*, ed. Earl Miner (Berkeley and Los Angeles, 1971).

Such micrography will be found sometimes in Jonson and becomes a compulsion in some of his Sons and their Sons. Herrick's frequent daintiness, Lovelace's preoccupation with the smaller animals, and Marvell's stress on the small (at least outside his public poetry)—all these show usage of the Anacreontic inheritance. When "lives short race" ends so soon, the Anacreontic shrug becomes a possible answer only on condition that the issues involved be viewed in reduced terms. The consolations of micrography would no doubt bring Lovelace's "Grasse-hopper" first to mind. But what we see on opening Stanley's *Poems* is "The Gloworme," a creature widely interpreted for emblematic reading. To Stanley's lover, it is a kindlier version of Waller's rose. To James Howell, it is "the old Emblem of true Friendship."[11] And its intermittent light flickers as well through a couple of Marvell's poems. Other images of the small often verge on the macabre. Thomas Randolph writes "Upon the losse of his little finger" as a *memento mori* for "The other members."[12] And in *Upon Appleton House*, Marvell observes himself observing:

> And through the Hazles thick espy
> The hatching *Thrastles* shining Eye.
>
> (531-32)

The great void is understood only by finding it in the grass.

> And now to the Abbyss I pass
> Of that unfathomable Grass,
> Where Men like Grashoppers appear,
> But Grashoppers are Gyants there:
> They, in their squeking Laugh, contemn
> Us as we walk more low then them:

[11] Howell, *Epistolae Ho-Elianae*, 11th edition (London, 1754), I. 6. lvi (to Thomas Ham), p. 295.
[12] *The Poems*, ed. G. Thorn-Drury (London, 1929), pp. 56-57.

> And, from the Precipices tall
> Of the green spir's, to us do call.   (369-76)

Marvell reminds us, if reminder be necessary, that altogether serious issues lurk in the fanciful. There is a shudder as well as a shrug in the Anacreontic conception of what I can only term the *bon vivant*. Somewhat less familiar examples of micrographic art will suggest that other poets than the famous M. P. for Hull, or his Mower, found significance in things of sizes not unlike "Glo-worms" or "one Blade of Grass." Lovelace has an emblematic "Ant."

> Forbear thou great good Husband, little Ant;
>   A little respite from thy flood of sweat;. . .
> Cease large example of wise thrift a while,
>   (For thy example is become our Law)
> And teach thy frowns a seasonable smile.
>
>   (1-2, 7-9)

The parenthesis suggests (as do passages in many of Lovelace's animal poems) a contemporary political relevance as hard to resist as to delineate. In a poem in which the moral Cato appears more than a little relaxed (1. 9) and in which Lucasta casts the sunshine of her eyes around the world, we may well ask about the significance of the Ant.

> Ant to work still; Age will Thee Truant call;
> And to save now, th' art worse than prodigal.
>
> *Austere* and *Cynick*! not one hour t'allow,
>   To lose wth pleasure what thou gotst with pain:
> But drive on sacred Festivals, thy Plow;
>   Tearing high-ways with thy ore charged Wain.
> Not all thy life time one poor Minute live,
> And thy o're labour'd Bulk with mirth relieve?
>
>   (17-24)

The Ant may seem at times a compulsive peasant, *animal laborans*; but the plowing on "sacred Festivals" certainly suggests agreement with the Parliamentary ordinance of 3 June 1647, "That the Feasts of the Nativity of Christ, *Easter* and *Whitsuntide* and all other festivall dayes commonly called holy dayes be no longer observed as Festivall or Holy dayes. . . ." Such "sacred Festivals" or holy days were not merely times of religious observation, as the Puritans well knew. The same ordinance makes this clear, saying that scholars, apprentices, and servants could *request* from their masters "such convenient reasonable recreation and relaxion from their constant and ordinary labours in every second Tuesday in the moneth throughout the yeare."[13] But the masters decided. And yet Lovelace does not simply turn his Ant into the Puritan; he talks more about the industriousness common to both. And we recall: "Go to the ant, thou sluggard; consider her ways, and be wise" (Proverbs, vi. 6). The difference of emphasis comes to this: for Lovelace, being wise involves, in the end, taking the Ant as a negative example.

> Look up then miserable Ant, and spie
> Thy fatal foes, for breaking of her Law,
> Hov'ring above thee, Madam, *Margaret Pie*,
> And her fierce Servant, Meagre, Sir *John Daw*.
>
> (33-36)

The Ant's labors are all lost, "Self and Storehouse," to its avian enemies. With the application of the emblem there comes another shift in the poem, however, to *us*. And with the final movement, we hear once again the Ovidian strains of Time the Devourer (Ovid's Age had appeared in l. 21).

[13] See William Hughes, *An Exact Abridgment of Publick Acts and Ordinances of Parliament* (London, 1657), p. 271.

> Thus we unthrifty thrive within Earths Tomb,
>   For some more rav'nous and ambitious Jaw:
> The *Grain* in th' *Ants*, the *Ants* in the *Pies* womb,
>   The *Pie* in th' *Hawks*, the *Hawks* ith' *Eagles* maw:
> So scattering to hord 'gainst a long Day,
> Thinking to save all, we cast all away.   (31-36)

The view of human life presented by the economy of small creatures does not really vary in Marvell's *To his Coy Mistress*.

> And now, like am'rous birds of prey,
> Rather at once our Time devour,
> Than languish in his slow-chapt pow'r.
>
>   (38-40)

Once again, we observe how the Anacreontic motifs present versions of the good life and its enemies such as we can find, in terms recognizably the same, but in very different imagery, in Herrick's *Corinna's going a Maying* and numerous other poems urging time as a "perswasion to enjoy."

Like many of Lovelace's best poems, "The Ant" refuses to be sorted out easily. A poem in which, within four lines, Cato watches naked actresses and Lucasta renders the day holy will long tarry on the race to simplicity. So much becomes clear: the grimness of life and the rejection of *extreme* diligence. The busy man may also go to Lovelace's Ant, consider its ways and be wise. He will not have reason to think art or life simple or happy; nor will any of us conclude that poetic micrography need necessarily be trivial. Such seemingly discordant elements combine in many Cavalier poems to make the small emblematic of the all. Such technique differs from the Donnean habit of retrieving the macrocosm on terms set by the microcosm; the Cavaliers chose rather to raise large issues from small details. We have seen as much in passages

from *Upon Appleton House*. It must be admitted that the
same poet's Mower declares that the small lacks meanings:

> Ye Country Comets, that portend
> No War, nor Prince's funeral,
> Shining unto no higher end
> Then to presage the Grasses fall. (5-8)

But by the end of the poem, we see that there can be disasters
in the country world as well as in the court world. The Mower
has followed the *ignis fatuus* (l. 12), and Juliana's "bright-
ness" has "displac'd" his mind. What may be ostentatiously
small may also be emblematically or otherwise crucial.

Whatever we make finally of the poise of subjects in Lovelace's consideration of the Ant, or whatever issues bear on
Marvell's "Glo-worms," we recognize another tone altogether
in a finer poem than either, Waller's "Of a fair Lady playing
with a Snake."[14] Waller begins by uniting his contrary worlds
in imagery more explicitly that of *discordia concors* than in
any of the micrographic Anacreontic poems so far considered.

> Strange that such Horror and such Grace
> Should dwell together in one place,
> A Furyes arm, an Angels face. (1-3)

Chloris "has a double guard": the snake and "her coldness."
Both images combine peculiarly in the last stanza but one.

> Contented in that Nest of Snow
> He lyes, as he his bliss did know,
> And to the wood no more would go.
> (13-15)

[14] In *The School of Love* (Princeton, 1964), a book most pertinent
to ch. v, below, H. M. Richmond reveals himself one of the few critics
to understand the merits of Waller's poetry. See pp. 144-45 for comment on this particular poem.

The erotic possibilities of the snake image are beautifully allowed and handsomely controlled by such imagery. Setting aside the merely sexual, one observes in the next stanza that her bosom, "that Nest of Snow," could warm a marble snake, so that Waller distinguishes sharply between what might be termed Chloris's chastity and the desire she arouses in the man. But it is by no means certain that she lives either properly or safely in her chastity.

> 'Tis innocence and youth which makes
> In *Cloris*'s fancy such mistakes,
> To start at Love, and play with Snakes.　(4-6)

The nice conjunction in the last line makes clear the amorous or erotic associations of the snake, although as amateur psychologists, most of us may have had amateur suspicions on encountering even the title. The suspicion is borne out in a somewhat eerie sensation felt by us as we read about the snake "boldly" creeping up Chloris's sleeve to her bosom. Here is a true Cavalier wit: to convey uneasiness by content, to portray innocence playing with sexual symbols, to convince us of the extent to which chastity may arouse, in Dr. Johnson's phrase, amorous propensities. But equally, youth's a stuff will not endure, and Chloris is on the move to the margin of sophistication. "To start at Love, and play with Snakes" has a great deal in common with "Worms shall try / That long preserv'd Virginity." It is time, ladies, time, and the good life is either to be gained or lost. The work and the snake are not wholly erotic; they are not wholly macabre. But elements of each quality combine to produce a sense of danger.

> Take heed (fair *Eve*) you do not make
> Another Tempter of this Snake,
> A marble one so warm'd would speak.　(16-18)

Five lines before this fear of the Fall of Eve, the snake was happy in its "Nest of Snow," and the male speaker had conveyed regret that "we dare not give / Our thought so unconfin'd a leave" as to follow the snake's progress through the young lady's clothes. The danger, then, had lain with the man, who found himself too amorously concerned. At the end, as at the beginning, however, the danger confronts Chloris: in time she may well prove to be Eve. If her mistakes involved but those of Innocence, and if she lived in a world without snakes, then she would be safe. But her mistakes also stem from youth, and the world is not simply pastoral. The good life and time as understood by what I have generalized as the Anacreontic conception in Cavalier poetry of course concerned other elements than Anacreon, and they will be our next concern. But such a conception, we may say as it were with Waller, possesses the wisdom of the serpent rather than any lasting innocence of the dove.

In moving to Horace as a representative of major elements in Cavalier poetry, we discover what seems very like our Anacreontic world: awareness of time and a degree of urgency in the enjoyment of life. But we also see major differences. The scale is much larger; not just one man grows old, but rather Time shares with Fate and Fortune a threat to a man in a world of equals. And the enjoyments sought are not quite the same. The search for the happy life is not undertaken by a *bon vivant* who sees death ahead and, with a shrug, asks for more wine. In short, more exists in Horace than Anacreon, and what exists is more complex. So much was recognized, and indeed Horace earned superlatives: "Horace. The best of Lyrick Poets."[15] There may be debate whether what is essen-

---

[15] Barten Holyday took the appellation (with a good deal else) from Sir Thomas Hawkins's translations. On Horace's extraordinary popularity from about 1580 to 1630, see my "Patterns of Stoicism in Thought and

tial to Donne's love lyrics should be best regarded as Petrarchan or anti-Petrarchan; but I believe that Cavalier poetry was, so to speak, pre-Petrarchan in its major emphasis. Every student of Cavalier poetry will recognize at least one phrase from *Odes*, I. xi. 7-8:

> dum loquimur, fugerit invida
> aetas: *carpe diem*, quam minimum credula postero.[16]

What students perhaps would overlook in the ode is the imagery of winters and the sea (ll. 4-6), although these images will prove of continuous importance in Cavalier and some later poetry. To move back to familiar ground, it is well known that *Epodes*, ii, "The praise of the Countrey life," as Hawkins put it, was central to poetry of retirement, its first two words, *Beatus ille*, having provided poets and their students before me with a great deal to consider about what I have called the happy life.

What is harder but more important to convey is that characteristic movement of Horace's odes from attention to public affairs, fortune, fate, Rome—things that make up the metaphysical, public, or social worlds—to a few friends, a single friend, or a mistress. Again and again, Horace moves from trouble, cold, or confusion without to security, warmth, and integrity within. He moves from public pressures to private relaxation with wine and love or to the strength of the whole man. He retreats from danger and greatness to more modest refuge in the storm. *Odes*, I. ix in fact gives such a movement

---

Prose Styles," *Publications of The Modern Language Association*, LXXXV (1970), 1023-1034, Table P.

[16] My italics isolate the phrase which of course means something like seize the day, use your hour. Hawkins is freer: "Whilst we are talking, Envious Time doth slide: / This day's thine owne, the next may be deni'de." His argument to the poem gives as reasons to enjoy life "the shortnesse of life, and speed of death."

twice (rather to the confusion of classicists) in lines 1-8 and
9-24: from winter outside to warmth and wine indoors; from
the waves in fortune's tempest to loves in corners of the Cam-
pus. The same spirit and movement mark III. viii and II. ix.
In the latter, roses, wine, and love are sovereign remedies for
the headaches induced (as it were) by the headlines. A care-
ful reader will find numerous other motifs and movements in
the *Odes*, and some of these may also claim stress. *Odes*, I.
iii, which bids farewell to Virgil setting out for Greece, meditates on the dangers of the sea and on man's folly in daring
too much. A great public ode to Augustus (I. ii) provided a
model for addresses to kings and princes in its basic form (as
late in our century as Dryden's *Britannia Rediviva*) and in its
formal properties (compare Marvell's *Horatian Ode*). Some
of the motifs of the poem are grown familiar: winter and
rising seas. Others will be found in Cavalier poetry: chaos
(fish in the trees) and confession of guilt. But someone wishing to embrace as much of Horace as possible in a single poem
surely should study *Odes*, III. xxix, the address to Maecenas
mentioned toward the end of the last chapter. There will be
found all the movements and motifs I have mentioned com-
bined into a splendid whole.

I have chosen to emphasize specific Horatian poems, with
their movements and motifs, because I see no other way to
emphasize the centrality of Horace to Cavalier poetry. Perhaps I have relied over much on assertion, but there remain
large tracts to cover, and rather than expect the reader either
to rely on my assertion or bear with lengthy quotations of
Horace, let me simply urge that he read the poems I have
mentioned and then the Cavaliers. And I think that my fellow students of Cavalier poetry will recognize a relevance in
the fact that a poet like Dryden, who was studying Horace at
Westminster in the former half of the century, should have
chosen in the latter half to translate (for *Sylvae*) precisely

those poems of common benefit to the Cavaliers: *Odes*, I. iii and ix; III. xxix; and *Epodes*, ii.[17] And what is pertinent in Dryden's understanding of Horace should be significant in the poetry of the Cavaliers themselves.

We may consider Lovelace again, to show how a poet given to Anacreontic gestures might find Horace important. His "Advice to my best Brother. Coll: Francis Lovelace," begins:

> *Frank*, wil't live handsomely? trust not too far
> Thy self to waving Seas, for what thy star
> Calculated by sure event must be,
> Look in the Glassy-epithite and see.   (1-4)

Fickle seas and Fate or Fortune—these we have observed repeatedly in Horace. So much is obvious. But what is that "Glassy-epithite"? The phrase is intelligible only by reference to the first stanza of *Odes*, IV. i, an ode on Pindar later imitated by Cowley. That is, one should look into a sea as calm as that sea known by its epithet, the Icarian, which for all its calmness is a warning against rash endeavor. How far this is clear may be judged from Horace's stanza and the story of Icarus; how far it is relevant can be judged by Lovelace's ensuing lines.

> Yet settle here your rest, and take your state,
> And in calm *Halcyon*'s nest ev'n build your Fate;
> Prethee lye down securely, *Frank*, and keep
> With as much no noyse the inconstant Deep
> As its Inhabitants.   (5-9)

The image of quiet nesting in the "inconstant Deep" appears altogether paradoxical, but it is well sanctioned by Horace and his interpreters of *Odes*, III. iii, to which was given the

---

[17] Also, as I have remarked, I. ii to Augustus underlies *Britannia Rediviva*; see *The Works of John Dryden* (Berkeley and Los Angeles, 1969), III, 474, 476, 479, 483.

motto, *mediis tranquillus in undis.*[18] It may well be that Francis Lovelace was planning at this time to sail for the New World and that the Horatian imagery came naturally to Lovelace's mind. If so, the very naturalness proves the importance. The last 38 lines (of the poem's 66) are closely based on yet another Horatian ode (ii. x), in which Horace, using once more the sailing imagery, advises Licinius to trim with conditions, propounding as the wise person that man, in Lovelace's phrase, "Who loves the golden mean" (l. 29), which is nothing other than a famous line or so from Horace, "auream quisquis mediocritatem / diligit" (ii. x. 5-6). Lovelace's poem comes close to being an "imitation" of Horace, as it would be termed in about four decades.

With such poems by Cavalier and classical poets in mind, and with the relation verified by attention to Horace and the Cavaliers apart from my assertion, we have come to a point where it is possible to generalize on the relation between time and the good life. We may take as our text a passage from Lovelace's poem to his brother.

> A breast of proof defies all Shocks of Fate,
> Fears in the best, hopes in the worser state;
> Heaven forbid that, as of old, Time ever
> Flourish'd in *Spring*, so contrary, now never:
> That mighty breath which blew foul Winter hither,
> Can eas'ly puffe it to a fairer weather.
> Why dost despair then, *Frank*? *Aeolus* has
> A *Zephyrus* as well as *Boreas*. (56-63)

We see a relation between time as accustomed process and

---

[18] The motto is attached to a picture illustrating the ode in Otto van Veen, *Q. Horatii Flacci Emblemata* (1607) and is reprinted by Maren-Sofie Røstvig, vol. 1, 2nd ed. (Oslo, 1962), opposite p. 26. In the first edition Professor Røstvig had associated the image with Lipsius's *De Constantia*. The association will also be found in Spenser.

the life good enough to endure: to it the spring returns. And we see a relation between time considered as the present state of England and the good life: character must triumph over time. Time may, then, be an ally, or it may be an enemy. More likely it is hostile. Winter, tempests, inconstant seas, or the northern wind represent a group of motifs inimical to the good life and associated with the advance of time, especially in the advance to one's own time and hour. On the other hand, wine and love are equally emblematic; we should not assume that the Cavaliers were talking merely about the bibulous and amorous. Such inclinations are defense measures, especially for the happy life in bad times. Similarly, the defiance of Fate by seeking out a calm haven or, on the contrary, by challenging Fate with one's breast, these two are motifs, now in the defenses of the good life. Lovelace's poem embodies both of these: "Draw all your Sails in quickly . . ." (ll. 72 ff.); "A breast of proof defies all Shocks of Fate" (l. 56).

One set of motifs will suggest that the happy life is being protected, the other the good life. It may not be realized how easily the two elements might be merged, however, or how often the pressure of time was discovered in the sorry state of the times. Two poems rather more explicit than those discussed so far may be given, at least in part, as evidence of the relation between the good life and time. Thomas Randolph's poem, "To Time," begins in hallowed Ovidian terms.

Why should we not accuse thee of a crime
And justly call thee envious Time;
When in our pleasures we desire to stay,
With swallows speed thou flyest away;
But if a griefe in our sad hearts doe keepe
Then thou art like a snail & wilt not creepe.

(1-6)

"Learn of Eternity not to change," he urges (l. 13), that is, while he is abed with his mistress. "But when the oft repeated acts of Love / Grow stale" (ll. 15-16) he feels differently: "shake off thy drowzy chaine / And gently Time take then thy wings againe" (ll. 23-24). Perhaps we may pause for a moment to reflect on Randolph's great reputation at the time. It has not lasted, and although there is much that is fine in what he writes, he is not a major Cavalier poet. One is inclined to follow the disingenuous example of Thomas Fuller and say of the brilliant Randolph:

> The Muses may seem not only to have smiled, but to have been tickled at his nativity, such the festivity of his poems of all sorts. But my declining age, being superannuated to meddle with such ludicrous matters, consigneth the censure and commendation of his poems . . . to younger pens, for whom it is most proper.[19]

Fuller's praise is not unrelated to the fact that he and Randolph came from the same county, and his declining to establish the reasons for praise is no doubt implied in the tone of the passage: there is a boyishness about Randolph, a freshness, that appeals greatly, to a point. We do not, however, find a greatness that appeals freshly. We must indeed recognize the possibility that one can write on time and the good life in poetry of varying quality.

Horace presents a different alternative. So does Lovelace in his advice to his brother, with one of the few examples of the strong line (Herrick provides others) to be found after Jonson. We believe Lovelace, as we believe Horace. So, too, do we believe Herrick, although he is too often classed among the pretty poets whom some critics have felt too "superannuated to meddle with." I promised a more explicit poem on

---

[19] Fuller, *The Worthies of England*, ed. John Freeman (London, 1952), p. 441.

the relation between the good life and time, and particularly a poem that treated time in terms of the ills of the time. My example will be, "The bad season makes the Poet sad." Herrick's so-to-speak implicities of the motifs of winter, love, and wine remain present, even in the explicit emergence of the speaker with his worries. All fourteen lines may be given.

Dull to my selfe, and almost dead to these
My many fresh and fragrant Mistresses:
Lost to all Musick now; since every thing
Puts on the semblance here of sorrowing.
Sick is the Land to'th'heart; and doth endure
More dangerous faintings by her desp'rate cure.
But if that golden Age wo'd come again,
And *Charles* here Rule, as he before did Raign;
If smooth and unperplext the Seasons were,
As when the *Sweet Maria* lived here:
I sho'd delight to have my Curles halfe drown'd
In *Tyrian Dewes*, and Head with Roses crown'd.
And once more yet (ere I am laid out dead)
Knock at a Starre with my exalted Head.

There is no need to explicate the poem. But most readers may wish to be reminded that the last line is a translation of the concluding line of Horace, *Odes*, I. i. And those who think that Herrick is otherwise unpolitical may wish to compare the last four lines with the frontispiece of the *Hesperides*, its great crown on the titlepage, and its motto, adapting Ovid (*Amores*, III. ix. 28), "Song alone escapes the greedy funeral pyre." We are ready to consider Time our certain enemy.

ii. The Ruins of Time

Surely every reader recalling the Cavalier poets thinks first of poems featuring transience or the pressure of time. It means one thing that Herrick has some forty poems in which

Time's arguments are heard, and it means another that those forty include his most familiar. To look at a few lines from one of them, "To Daffadills," we see from the very structure of the two stanzas how much the argument to the flowers argues to man.

> Faire Daffadills, we weep to see
>   You haste away so soone:
> As yet the early-rising Sun
>   Has not attain'd his Noone.
>   Stay, stay, . . .
>
> We have short time to stay, as you,
>   We have as short a Spring;
> As quick a growth to meet Decay,
>   As you, or any thing.
>   We die, . . .   (1-5, 11-15)

Time is the agency imparting to Cavalier lyrics that golden sadness which makes them the *fragmenta aurea* of every cultivated reader.

> Come, my *Corinna* come; and comming, marke
> How each field turns a street; each street a Parke
>   Made green, and trimm'd with trees: see how
>   Devotion gives each House a Bough,
>   Or Branch: Each Porch, each doore, ere this,
>   An Arke a Tabernacle is
> Made up of white-thorn neatly enterwove;
> As if here were those cooler shades of love.
>   Can such delights be in the street,
>   And open fields, and we not see't?
>   Come, we'll abroad; and let's obay
>   The Proclamation made for May:
> And sin no more, as we have done, by staying;
> But my *Corinna*, come, let's goe a Maying.   (29-42)

Perhaps Herrick employs something of the antipathetic fallacy dear to the Petrarchans: all else flowers but my love, because she will not grant hers. But the urgency resides in the fear that what we have presently we shall not have anon. Indeed, more than the usual pressure of time can be discovered in the poem. Although we cannot be sure what Herrick meant by "The Proclamation made for May," and although I do not know when the poem was written, its "Proclamation made for May" was null and void when it was published in 1648. The "sad times" or the "bad season" as it is variously termed by Cavaliers had come, and the Maypoles were put away.

> Come, let us goe, while we are in our prime;
> And take the harmlesse follie of the time.
> We shall grow old apace, and die
> Before we know our liberty.

One cannot claim that these lines offer political prophecy. But like all good poetry they possess this much of prophecy: the capacity to acquire relevance *with time*.

The tone of Cavalier poetry varies considerably, and Herrick's "Fresh-quilted colours" will not be found in all such poems. Yet his argument, or Time's argument, echoes and reechoes in Cavalier poetry.

> But at my back I alwaies hear
> Times winged Charriot hurrying near:
> And yonder all before us lye
> Desarts of vast Eternity.
> Thy Beauty shall no more be found,
> Nor, in thy marble Vault, shall sound
> My ecchoing Song.   (21-27)

Marvell's Coy Mistress must confront (with far less in the way of colors to tint the reality) the human condition: what life holds that is good must be taken quickly, because it will soon be gone. The arguments of Time's ruins sometimes seem to comprise the Cavaliers' arts of logic and rhetoric together, because Time's reasons are meant to persuade. For example, like Campion before him, Jonson (as we have seen) could recall Catullus:

> Come my *Celia*, let us prove,
> While we may, the sports of love;
> Time will not be ours, for ever.[20]

The most explicit argument—"Time will not be ours, for ever"—will not be found in Catullus v., although the thought of course occupies Catullus's mind enough to use the argument of death's long night to Lesbia. So far do the Cavalier poets feel Time at their backs or starkly before them that the villain must be branded by name, as in one of the most Roman of Jonson's love elegies.

> As time stands not still,
> I knowe no beautie, nor no youth that will.
> To use the present, then, is not abuse.[21]

Carew's poem, "To A. L. Perswasions to Love," some 85 lines long, deserves study and respect. It lacks the movement of Herrick, or Waller, or Marvell at their best, but it manages to be more immediate. Invitations to go a-Maying, telling lovely roses to go, and talk of hypothetical situations in which the lady might be loved with great ceremony—none of these

[20] From Jonson, *Volpone*, III. vii., where the situation is hardly that of Catullus's youthful air with Lesbia. But the poem was also printed in *The Forrest* and set to music as a song too good to be left to that miserly lecher, Volpone. On the urgency behind Marvell's *To his Coy Mistress*, see Stanley Stewart, "Marvell and the *Ars Moriendi*," *Seventeenth-Century Imagery*.

[21] Jonson, "An Elegie" ("By those bright Eyes"), ll. 19-21. The situation is Ovidian in its impatience and its adulterous character.

marks Carew's argument. His is too urgent to be directed *ad floram*.

> Did the thing for which I sue
> Onely concern my selfe not you,
> Were men so fram'd as they alone
> Reap'd all the pleasure, women none,
> Then had you reason to be scant.   (17-21)

Unreality is rejected for being unreal, and the immediacy lends real vivacity to the reality of Time's argument. We believe Carew at once.

> Oh love me then, and now begin it,
> Let us not loose this present minute:
> For time and age will worke that wrack
> Which time or age shall ne'er call backe.
>
> (69-72)

Only at the end (we recall) does Carew permit himself the floral image, and at that stage it exerts renewed force.

> Oh, then be wise, and whilst your season
> Affords you dayes for sport, doe reason;
> Spend not in vaine your lives short houre,
> But crop in time your beauties flower:
> Which will away, and doth together
> Both bud, and fade, both blow and wither.
>
> (79-84)

We are dying animals. That simple reality is the one confronted by Cavalier poets treating love and other subjects. And even if the lady yields, time's lesson remains the same, as Lovelace shows in his "Song. To Amarantha."

> Heere wee'l strippe and coole our fire
> In Creame below, in milke-baths higher:

> And when all Well's are drawne dry,
>> I'le drink a teare out of thine eye.　　(21-24)

Something has been gained. And something will be perforce be lost.

>> Which our very Joyes shall leave
>> That sorrowes thus we can deceive;
>> Or our very sorrowes weepe,
>> *That joyes so ripe, so little keepe.*
>>> (25-28)

The argument turns out to be stronger than the purpose for which it was designed. And so, too, when the lady does not yield:

>> Then die, that she,
>> The common fate of all things rare,
>> May read in thee.

Not all poetry seems to believe so radically that Time is the prime mover, nor need all poetry concern itself with that belief, with that aspect of human life. The conviction that Time and Age and Change work always and work everywhere their ravages transcends the desire to find the happy life by seizing the day. Above all, the ruins of time are not an idea, abstract and cold. Mutability, even death, can be *seen* in the flower, Waller implies; and Marvell says that Time's chariot can be *heard*. In *The Vision of Delight*, Jonson's choir goes yet farther.

>> We see, we heare, we feele, we taste,
>> we smell the change in every flowre,
>> we onely wish that all could last,
>> and be as new still as the houre.　　(136-39)

Surely we had not required such explicitness. Surely that sense of man's little while that animates Chinese and Japanese literature as well as our own has grown timely, and it is no longer bad manners to say that we will shortly die? I cannot believe the lovely close of *Corinna's going a Maying* anything other than a most suitable expression of that unsuitable fact.

> So when or you or I are made
> A fable, song, or fleeting shade;
> All love, all liking, all delight
> Lies drown'd with us in endlesse night.
> Then while time serves, and we are but decaying;
> Come, my *Corinna* come, let's goe a Maying.[22]

Like Jonson, Herrick has a vision of delight—"All love, all liking, all delight"—so soon and so endlessly to be lost. How good life must be, if the loss is so grievous. And how universal are Time's workings. By the end of Herrick's poem, the "or you or I" becomes "we." At our best, "we are but decaying." Another poem, "The Changes to Corinna" lacks the coloring of the famous poem, but it certainly has its argument in little.

> You are young, but must be old,
> And, to these, ye must be told,
> Time, ere long, will come and plow
> Loathed Furrowes in your brow:
> And the dimnesse of your eye
> Will no other thing imply,
> But you must die
> As well as I.   (11-18)

The cruel arguments of Time seem unanswerable. And perhaps a glance or two at poems less familiar to most readers will show with how much conviction those arguments were

[22] Herrick, ll. 65-70. This is a good point at which to stress the virtues of J. Max Patrick's edition of Herrick, *The Complete Poetry of Robert Herrick*, 2nd ed. (New York, 1968). See, for example, his glossing of these lines from The Wisdom of Solomon, ii. 1-8 and Proverbs, vii. 18.

harkened to. Herrick wrote, for example, a brief poem, "To his Kinswoman, Mistresse Susanna Herrick."

> When I consider (Dearest) thou dost stay
> But here awhile, to languish and decay;
> Like to these Garden-glories, which here be
> The Flowrie-sweet resemblances of Thee:
> With griefe of heart, methinks, I thus doe cry,
> Wo'd thou hast ne'r been born, or might'st not die.

No one would seriously argue, having read this poem, that it rates esteem above *Corinna's going a Maying*. At the same time, however, after reading this poem addressed to a beloved young lady relative, I do not see how we can deny to the poems of greater stature an equally felt, vital concern with time and mutability and age. The cry from the heart for Susanna Herrick may often be made, by greater art, into songs addressed to Celias, Lucastas, and Corinnas. He who argues that art transforms or transcends life will get no argument from me. But by the same token, there is only life to transform.

Herrick's sense that "all human things are so frail and uncertain," as Cicero may be more closely rendered, is revealed in one of his best but little known poems, "A Paranaeticall, or Advisive Verse, to his friend, Master John Wicks."[23] The first four words pose the problem: "Is this a life . . . ?" What life should we then desire? Not the laborious toil of Lovelace's Ant, but ample provision. And

> A Pleasing Wife, that by thy side
> Lies softly panting like a Bride.

---

[23] Wicks, or Wickes, or Weeks, was ordained with Herrick in 1623. Herrick wrote another fine poem to his friend, "His age" (Patrick ed., pp. 179 ff.), but to my taste, the classical elements are less wholly absorbed there and the feelings less finely expressed.

> This is to live, and to endeere
> Those minutes, Time has lent us here.
>
> (13-16)

Then come Time's emblems.

> Time steals away like to a stream,
> And we glide hence away with them.
> *No sound recalls the houres once fled,*
> *Or Roses, being withered:*
> Nor us (my Friend) when we are lost,
> Like to a Deaw, or melted Frost. (22-27)

We cannot escape the conviction that Time's best argument is death. But more remains to be said. The lines immediately following introduce a subject that will occupy me particularly in the next chapter: another version of Time, that is, the times.

> Then live we mirthfull, while we should,
> And turn the iron Age to Gold. (28-29)

It will also be observed that a certain hope springs from those two lines. That hope, and the resolution with which death is regarded in the conclusion of the poem, prepare the way for my next concern, how to resist time.

> *Whose life with care is overcast,*
> *That man's not said to live, but last:*
> *Nor is't a life, seven yeares to tell,*
> *But for to live that half seven well:*
> And that wee'l do; as men, who know,
> Some few sands spent, we hence must go,
> Both to be blended in the Urn,
> From whence there's never a return. (32-39)

We recall those strains of Jonsonian music about the good life: "It is not growing like a tree . . ."; "A Lillie of a Day

... and flowre of light." Man could find within himself measures enabling him to face his enemy, Time.

iii. The Remedies of Time

One, or perhaps several, of Ovid's charms will be found in his ability to stand on both sides of an issue. After writing *The Art of Love*, he wrote *The Remedies of Love*, so that, as he put it (1. 44), the same hand that wounds will bring the cure. If there are *remedia amoris*, time also required for the Cavaliers its remedies: cures, if possible and, if not, then lenitives, balms, and assuagements. And although my business in the rest of the chapter lies with investigation of the Cavalier *remedia temporis*, I hope we may spare a moment to recall a few of the great prose writers contemporary with the poets. The recurrent religious crises of the century brought many people those severe doubts called cases of conscience. The spell of Jeremy Taylor's magical style, and the experience of life that informs his advice, still keep alive his "Paranaeticall, or Advisive" works, his two rules and exercises, *Of Holy Living* and (much the more popular) *Of Holy Dying*. And in *The Anatomy of Melancholy* (in which one will not necessarily find what one recalls but is sure to find much else), Robert Burton offers (II. iii. i. 1) a consolation ". . . containing the Remedies of all manner of Discontents." And perhaps my reader will allow me to say with Burton that "I have thought fit, in this following Section, a little to digress, (if at least it be to digress in this subject)," and to ask of Herrick and Jonson, or of Taylor and Burton, "To what end are such paranaeticall discourses?" Are there, after all, remedies of our problems? Well, "Whatsoever is under the Moon is subject to corruption, alteration; and, so long as thou livest upon the earth, look not for other." Burton's map of life indicates some delectable vales: "If the way be troublesome, and you in

133

misery, in many grievances, on the other side you have many pleasant sports, objects, sweet smells, delightsome tastes, musick, meats, herbs, flowers, &c. to recreate your senses." And on the same map one will find, at the end, beyond the painful terrain, the delectable mountains: "our life is a warfare, and who knows it not? . . . Go on then merrily to heaven."

Cavalier poets do not sing Burton's merry songs, and they do not offer anything resembling a transcendental view of life: so goes the conventional wisdom. But it does not go very far. More truly, we will not find them gazing raptly at eternity the other night, and equally truly *some* of their remedies of time are not transcendental. But in the end, we shall have to conclude that their poetry possesses its own kind of transcendence, not of the kind that flies clear out of this world, but of the kind in which individual character rises above time, fortune, and adversity. The difficulty in setting such matters straight derives partly from the fact that the religious poetry of, say, Herrick is more ceremonial than devotional, and partly from the fact that Time was a metaphysical, almost a transcendental enemy, whereas man occupied an existential position in time. It almost seems that Time drives his iron wedges between eternity and man, placing man in the agonized position of Marvell's speaker in *To his Coy Mistress*. And yet, if that existential dilemma does not partake of a kind of transcendence, one is very much mistaken. We may begin, then, to consider certain remedies of time, beginning with the simpler.

The Cavalier poets of course knew their Ovid, but the changes of forms so crucial to the *Metamorphoses* alter in the transformations that the Cavaliers sometimes use to defeat Time. The difference needs no great emphasis. Escapes from time and change by one last, major change were not really open to Cavalier poets, and they probably would not have sought them if they could have. For one thing, none of them seems to have felt with Ovid that the human epic is so essentially tragic,[24] or that there were no other remedies. For another, Ovid did not suit with Christianity at quite this juncture. But having said all these things, one must face up to the fact that Ovid is the most popular of all poets in England throughout the century, that his vision of Time the Devourer of Things is the commonplace, and that he could still teach poets lessons that he had not intended. Herrick wrote a number of "How" poems, explaining how this or that came to be what it is. Such a concern with what changes have been involved in things becoming what they are certainly will be found in Ovid. Herrick's "How Springs came first" may serve as example.

> These Springs were Maidens once that lov'd,
> But lost to that they most approv'd:
> My Story tells, by Love they were
> Turn'd to these Springs, which wee see here:
> The pretty whimpering that they make,
> When of the Banks their leave they take;
> Tels ye but this, they are the same,
> In nothing chang'd but in their name.

Here is a remedy of Time, here the Ovidian change. But like almost all of Herrick's other poems of metamorphosis, and unlike his poems on time, this "pretty whimpering" carries little conviction. Pleasing finger-exercises do not serve as real remedies of time. Much the same dismissal must ultimately be made of Carew's charming little poem, "Upon a Mole in Celias bosome."

Carew's song, "Aske me no more," presents an altogether finer version of remedy by metamorphosis. He begins with the very emblem of transient beauty, the rose.

[24] See Brooks Otis, *Ovid as Epic Poet* (Cambridge, 1966).

135

> Aske me no more where *Jove bestowes*,
> When *June* is past, the fading rose:
> For in your beauties orient deepe,
> These flowers as in their causes, sleepe.   (1-4)

The situation implies that for once the woman hears and is troubled by Time's arguments, and that for once the man bears conviction that she is the abiding principle. The natural phenomena change their form by returning to their Aristotelean causes, to "sleepe" for rebirth in the "deepe" beauty of the woman. Earlier in this chapter I emphasized the importance of the Horatian movement from outside to within, from the bustle of the many to the friendship of the few—and other such directions. I think that it will be seen that such a movement is crucial to Carew's poem, crucial also to metamorphosis as a remedy of time, and crucial finally to forms of transcendence in Cavalier poetry. Carew did not quite see Blake's "Heaven in a Wild Flower," nor did he hold "Eternity in an hour." But he does something very like this, as his last stanza shows.

> Aske me no more if East or West,
> The Phenix builds her spicy nest:
> For unto you at last she flies,
> And in your fragrant bosome dyes.   (17-20)

The phoenix, dying only to be reborn as the single example of its species, provides an image accommodating change and permanency, and once again renewal becomes possible by touching the abiding cause, the lovely woman. Some very subtle things are happening all this while. Just as the rose in the first stanza had by tradition been an emblem for a woman's beauty and its transience, providing a certain leakage of the usual tenor of the image into the new meaning, so with

136

the phoenix: the woman addressed is phoenix-like, unique, and to that extent no comfort at all to other women, or perhaps to the speaker as he contemplates time. She gains reassurance that the signs of transience all about her (roses, sunbeams, the nightingale of May, falling stars, the phoenix) provide the proof of her permanence as cause. The rest of the world does not share her comfort, however. The reason for her triumph, for her becoming as it were the stable vessel into which changes could come for cyclic renewal, is what might almost be called her saintly excess of merit. Only one possessing to so high a degree the merit she does, that is, beauty, can retain herself as herself and indeed as a renewal of others. We shall see repeatedly how this Horatian direction to the blessed few marks the transcending process in Cavalier poetry. The remedies of time are not allotted to Selden's "Most." We recall that at the end of the *Metamorphoses*, after the moving depiction of change in Book xv, Julius and Augustus Caesar are granted an apotheosis denied other men.

The more public poetry of the Cavalier poets often chose to follow Ovid in apotheosis as a remedy of Time. Waller's *Upon St. James's Park* may be recalled, or Marvell's *A Poem upon the Death of O. C.* might be examined. But surely we have reached the point where but one poem can be allowed, either by tradition or by present need, Jonson's *To the Memory of My Beloved, the Author Mr. William Shakespeare: And What He Hath Left Us*. By examining it as a poem concerned with remedies of time and especially of apotheosis as triumph over our enemy, Time, I think we may find fresh appreciation for a familiar poem, and some unfamiliar kinds of transcendence allowable to Jonson. Jonson begins in a way that readers have always found difficult to assess. He starts inquiringly, problematically, apologizing for writing so "ample" a poem,

While I confess thy writings to be such,
    As neither *Man,* nor *Muse,* can praise too much.
                                                    (3-4)

As himself man and Muse (in the common sense of a poet),
Jonson fends off others who would praise Shakespeare inade-
quately. To Jonson, praise by the unqualified ranks as dis-
praise. All this may be true, but it seems a coming through the
door sideways when it is wide open. Or perhaps a truer comparison would be to a great general slowly preparing for de-
cisive victory.

Jonson's first sudden movement, and it does come with an
effect of suddenness after the slow beginning, turns away from
praise of the man in his "writings" to identification of the
man as the very *Zeitgeist.*

I, therefore will begin. Soule of the Age!
    The applause! delight! the wonder of the Age!
    My *Shakespeare,* rise . . .   (17-19)

That is pure Jonson; praise of the greatest of his kind, along
with sturdy attention to himself. We see even this soon that,
as with Carew's lady, so with Jonson's Shakespeare, the finest
of his kind is the motive cause ("Soule") of the age. But with
this difference: he lives only when another great poet summons him. The dead shall be raised, only because they are
summoned—and because they are not really dead: Shakespeare is "alive still, while thy Booke doth live" (23). In this
he differs from those dramatists his contemporaries (25-30).
For mettle like his, one must compare him to the Greek and
Roman dramatists, whom he also excels (31-40), even as he
surpasses "all Scenes of *Europe*" (41-42). Shakespeare therefore "was not of an age, but for all time!" The "Soule of the
Age" is soul of all ages.

Such immortality in the classical sense (of one born in time but lasting throughout time by his achievements) leads to alternative comparisons. Shakespeare "like *Apollo* . . . came forth to warme / Our eares, or like a *Mercury* to charme" (45-46). Shakespeare now has taken on qualities of poetry (also the sun) and music (also magic) in godlike fashion. Such godlikeness renders all he does natural (44-54) and yet artistic (55-70). What more could one achieve?

Jonson seems to have a problem, and to some readers he may seem to acknowledge rather than to solve it by his very familiar apostrophe, "Sweet Swan of *Avon!*" (71). So familiar is the phrase that we usually fail to observe the crucial nature of the moment of metamorphosis. Shakespeare is not said to be *like* the Swan as he was "like *Apollo*" or "like a *Mercury*." He *is* the "Sweet Swan of *Avon*" to whom others might be compared. Perhaps the mythical and biographical, even the topographical, justice allows for the identification. Perhaps we accept it because it comes at the beginning of a peroration. Perhaps we accept it as being a lesser status than "Soule of the Age," or perhaps because it is familiar. It may seem that I make heavy weather of four famous words. But in very fact, Jonson has acknowledged at last—and for the first time in the poem—that Shakespeare is dead. The swan sings as it dies. And after those four words we read

> what a sight it were
> To see thee in our waters yet appeare,
> And make those flights upon the bankes of *Thames*,
> That so did take *Eliza*, and our *James*!   (71-74)

(Did Waller's "overhead . . . fowl" spring from Jonson's lovely conceit?) The tenses and the situation tell us that the Swan is dead, his poetic "flights" done. The movement of the poem surely must seem strange, to this point, from deathless immortality to death. "But stay," exclaims the poet, "I see thee in the *Hemisphere* / Advanc'd, and made a Constellation there!" (75-76). The metamorphosis into a swan, with all the accompanying complexity of tone and poetic movement, turns out to have been but preparation for another metamorphosis into a swan, that is, into the swan-constellation, into Cygnus, the "Starre of *Poets*" (77). The apotheosis resembles that furnished by Ovid for the first two Caesars, but Jonson takes pains to Christianize Cygnus by showing that Shakespeare has become the intelligence, as it was called, the saintly soul intermediate between divine providence and decree on the one hand and man on the other: "Shine forth, thou Starre of *Poets*, and with rage, / Or influence, chide, or cheere the drooping Stage" (77-78).

Jonson no doubt allowed himself such "ample" movement in his poem to make his apotheosis work. It must not be forgot that such transformation of Shakespeare into the "Starre of *Poets*" (a lovely phrase) appropriates to a mere actor-playwright the mythologizing imagery formerly thought appropriate only to princes, great heroes, and poets of respectable genres. Jonson succeeds without our even being aware of his problems. We think all along that his problem is one of tone, that is, of attitude toward Shakespeare and of creating a credibility of conviction. That was not the problem at all. What required doing was making a social nobody into what he truly was, a living influence on mankind ever since. Jonson's success can be gauged by comparison with another of his poems, that "Ode. ἀλληγορικὴ" to Hugh Holland. The "Ode" does not lack interest, but the darkness of its learned allegory and the sense of removal from us created by its masquelike qualities do not achieve the immediacy of the poem on Shakespeare.[25] In *that* triumphant elegy turned panegyric (and

[25] Jonson's ode was prefixed to Hugh Holland's *Pancharis* (1603), a poem directed by Holland to James I but much concerned with Owen Tudor and the ancient lore of the West. Holland's learning and the poem's regal associations allow him a similar apotheosis (ll. 5, 97 ff.).

surely Dryden owed more than he indicated to this poem), Jonson makes a complex process of metamorphosis / apotheosis work out in details that seem altogether historical and even almost realistic. The apotheosis takes place in time—in an "Age" and in the reigns of Elizabeth and James. But it also overcomes time, "for all time."

The fact and the mode of overcoming time are probably not as obvious as I have made them seem. The subtleties possessed by a great poem like Jonson's may forever elude explication and forever be felt by sensitive readers. But what "metamorphosis" and "apotheosis" imply may vary considerably without, however, forfeiting a basic characteristic, the altering of a state subject to time to one superior and usually immune to time. The mode remains constant, although the form of change may differ. The readiest illustration of such processes will be found in Jonson's masques. *The Gypsies Metamorphosed*, for three centuries a favorite among readers of the genre, turns on metamorphosis of the gypsies into members of the court. What was partial, reprehensible, and chaotic becomes whole, admirable, and orderly. The kind of alteration in other masques varies within the loose sense of "metamorphosis." Another rather simple example is provided by *The Masque of Blackness*, in which the daughters of Niger are promised change. *The Golden Age Restored* presents a less easily defined change. In a typical Jonsonian use of the antimasque, the Iron Age calls forth two devils that dance in a symbolic disorder. As so often, the change comes with something very like a *dea ex machina*: Pallas enters, turns the devils into statues, and calls in Astraea and the Golden Age. To normal usage, the change of devils into statues is a metamorphosis, but their replacement by the goddess of justice and the age of gold seems to deserve another term. The kind of alteration enacted seems to involve not so much the change of a single thing into another as a kind of displacement.

Similarly, Shakespeare's transformation into a swan and then into Cygnus would seem to differ from replacing Shakespeare by a swan or by the constellation.

The process of alteration by substitution may be seen at its simplest, and almost its finest, in Jonson's *Masque of Queens*. It begins with the antimasque, in which witches dance to "a strange and sodayne Musique . . . a *magicall Daunce*, full of praeposterous change, and gesticulation." Evil disorder parodies order in one of its symbols, the dance. But the chaotic antimasque yields to the House of Fame and Perseus at the "sound of loud Musique," at which the hags and their Hell mouth disappear. The stage direction gets us directly to the heart of our matter: "and the whole face of the *Scene* alterd; scarse suffring the memory of any such thing."[26] In the old dramatic terms, we observe a peripeteia, or reversal; and our observation involves the reader's version of an anagnorisis, a recognition. The change itself constitutes a kind of displacement, whereas the reader's understanding undergoes the metamorphosis. That is, he observes a change or substitution that involves his values. As a standard Greek-Latin dictionary put it, "Things are said to be truly reversed that are fallen in a contrary posture."[27] The tragic movement goes downward, but the movement we are considering turns upward to a new state: one either transcending or grown impervious to time. Such movement may include the Horatian centripetal transcendence or it may involve "The Golden Age Restored." Whether moving in this or the other direction, the metamorphosis, if the looser word may be allowed as the generic term, defeats time. Jonson's transformation of Shakespeare into the "Sweet Swan of *Avon*" seems mere metamorphosis,

[26] For the stage directions in *The Masque of Queens*, see *Ben Jonson*, VII, 301.

[27] "Res vero περιπετεῖς dicuntur quae in contrarium statum sunt lapsae": *Cornellii Schrevelli Lexicon* (etc.), 13th ed. (London, 1762), s. v. περιπετής.

except that the force of the poetry allows for a process of immortalizing. But the further alteration to the constellation Cygnus is so radical in its sequence from poet to animal to "Starre" as to make distinctions between metamorphosis and substitution seem Scholastic. When Jonson's masques so frequently depict alteration of one thing into another form, or substitution of one thing by another, or reconciliation of opposites (e.g., *Pleasure Reconciled to Virtue*), or return of what once was (*The Golden Age Restored*)—then it scarcely seems amiss to steal a leaf from Ovid and to speak of metamorphosis generically as a species of triumph over time.

Metamorphosis as a remedy of time may, then, move reductively to something dainty (the mole on Celia's bosom, for example), and it may engross something larger or finer (the restoration of the golden age). It is a tribute to Jonson's art that he seems to move to something smaller in the "Sweet Swan of *Avon*" in order to move to the larger constellation immutable beyond the sphere of the inconstant moon. The processes may well vary, but they are directed to the end of defeating time. Ovid almost concludes the last book of the *Metamorphoses* with the apotheoses of the Caesars. Almost: because he remains conscious that his poem is a poem and that he is a poet. At the end, it is the power of poetry, of what we may distinguish as a third remedy of time, the immortalizing power of art, a capacity that Jonson also believed he possessed.

The powers of the poet to confer upon his subjects an immortality beyond time had been prized by the Elizabethan poets, [28] but the *locus classicus* is the ode with which Horace concluded his first three books of *Odes* (III. xxx) when they went forth in 23 B.C.

[28] See J. W. Lever, *The Elizabethan Love Sonnet* (London, 1956), and O. B. Hardison, *The Enduring Monument* (Chapel Hill, 1962), two excellent books.

A Monument by me is brought to passe,
Out-living royall Pyramids, or Brasse,
Which neither shall consuming rayne abate,
Nor force to No[r]thern Tempests ruinate:
Nor yeares (though numberless): Nor Times swift start.[29]

Horace knew in advance of the enemies of poetry and of the
Cavalier vision of life: the winter rain, the north wind, the
years, and above all, the flight of time. Horace also provided
the Cavalier poets with what was probably their strongest
poetic faith, a faith in their art. Just as Pope was to be the
English Horace of the eighteenth century for his epistolary
and satiric verse, so Jonson was an earlier English Horace for
his epistles and odes. One difference between Jonson and
Pope, as also between Cavalier and Augustan poetry, will be
found by comparing Pope's great epistle to Dr. Arbuthnot, an
apologia for a satiric career, with Jonson's moving epistle "To
Elizabeth Countesse of Rutland."

Beautie, I know, is good, and bloud is more;
   Riches thought most: But, *Madame*, thinke what store
The world hath seene, which all these had in trust,
   And now lye lost in their forgotten dust.
It is the *Muse*, alone, can raise to heaven,
   And, at her strong armes end, hold up, and even,

[29] The rather inadequate version is by Hawkins, *Odes of Horace*
(London, 1635). Horace wrote

> Exegi monumentum aere perennius
> regalique situ pyramidum altius,
> quod non imber edax, non Aquilo impotens
> possit diruere aut innumerabilis
> annorum series et fuga temporum.

Ovid's claim at the end of the *Metamorphoses* (xv. 871-79) is yet more
remarkable, in view of the powerful sense of time and mutability
throughout the work as a whole.

The soules, shee loves. Those other glorious notes,
　　Inscrib'd in touch or marble, or the cotes
Painted, or carv'd upon our great-mens tombs,
　　Or in their windowes; doe but prove the wombs,
That bred them, graves: when they were borne, they di'd,
　　That had no *Muse* to make their fame abide.
How many equall with the *Argive* Queene,
　　Have beautie knowne, yet none so famous seene? . . .

　　　　　　　　　　　　　　　　　　　(37-50)

Here are a number of the crucial Horatian images and here
too an *O altitudo!* on the power of poetry that puts Jonson in
his century rather than the next.

How many equall with the *Argive* Queene,
　　Have beautie knowne, yet none so famous seene?
Achilles was not the first, that valiant was,
　　Or, in an armies head, that lockt in brasse,
Gave killing strokes. There were brave men, before
　　Ajax, or *Idomen*, or all the store,
That *Homer* brought to *Troy*; yet none so live:
　　Because they lack'd the sacred pen, could give
Like life unto 'hem. Who heav'd *Hercules*
　　Unto the starres? or the *Tyndarides*?
Who placed *Jasons Argo* in the skie?
　　Or set bright *Ariadnes* crowne so high?
Who made a lampe of *Berenices* hayre?
　　Or lifted *Cassiopea* in her chayre?
But onely *Poets*, rapt with rage divine?　　(49-63)

Shakespeare's apotheosis was one we had thought due only
to his own achievement in poetry; it now seems that Jonson's
art also played an effective role.
　　The poem to the Countess of Rutland surely arouses our

own conviction that Jonson's trust in the powers of art was a settled faith. That faith, shared by him with others, was soon to be tried, and some of the uncertainties of a revolutionary generation were to lead them to other remedies of time. And yet Jonson's example could not die. In his funeral elegy, John Taylor, "The Water Poet," wrote of *"a Metamorphosis* most strange" with Jonson's death, and in the collection, *Jonsonus Virbius,* William Habington revealed that Jonson had been apotheosized to the Pleiades, continuing his earthly role as pilot of poets.[30] As might be expected, the Horatian motif of raising immortal art appears with some frequency in *Jonsonus Virbius.* Falkland, for example, uses it in his pastoral elegy, and so of course does Sir Thomas Hawkins, the translator of Horace.[31] George Fortescue managed to be most explicit of all in his brief address "To the Immortalitie of My Learned Friend, M. Johnson." Jonson's "rage divine," as he put it himself in his epistle to the Countess of Rutland, has now become a power resembling that of God Himself, who may grant eternal life.

> Onely Ile say, Heaven gave unto Thy *Pen*
> A *Sacred power, Immortallizing* men,
> And thou dispensing Life immortally,
> Do'st now but *sabbatize* from worke, not dye.[32]

Jonson's admirers in 1638 returned to him the faith he had given his age: poetry, he said, could conquer time; his poetry, they said, had conquered time. Almost every poet and po-

---

[30] *Ben Jonson,* ed. C. Herford, Percy and Evelyn Simpson, 11 vols. (Oxford, 1925-1952), XI: The Water Poet, p. 425, ll. 155 ff.; Habington, p. 447, ll. 45 ff.

[31] Falkland in *Ben Jonson,* XI, 435, l. 240; Hawkins, p. 440, ll. 27-30; so also the rather touching poem by Shackerley Marmion, "A Funerall sacrifice, to the sacred memory of his thrice honoured Father Ben. Johnson," pp. 463-65.

[32] *Ben Jonson,* XI, 446, ll. 13-16.

etaster in England rushed to get a tribute into Falkland's collection, partly to get themselves associated with Falkland no doubt, but also to bask in Jonson's immortal sunshine. The political weather soon worsened, however, with the approach of civil war, the execution of Charles I, the triumph of Cromwell and, what hit the Cavaliers hardest in some ways, sequestration of property, punitive taxes, and Cromwell's major-generals. Like Horace's enduring monument, the Cavalier view of life was subjected to the devouring winter rain and the north wind. The hard times, as opposed to time, however, will be a concern of the next chapter. It remains to say that one of Jonson's Sons, Robert Herrick, found with Father Ben and with Horace, Jonson's Roman ancestor on his mother's side, as Randolph put it, that he, too, could raise a poetic monument defying time.

> Behold this living stone,
> I reare for me,
> Ne'r to be thrown
> Downe, envious Time by thee.
>
> Pillars let some set up,
> (If so they please)
> Here is my hope,
> And my *Pyramides*.[33]

The Horatian and Ovidian echoes are obvious enough, although felicities like "living stone" may be too quickly passed. The whole poem is one of Herrick's best in at least the sense that it drives into areas of experience that readers of anthologies have not found him in. The second and the fourth stanzas tell us of his and our common enemy, Time.

---

[33] Herrick, "His Poetrie his Pillar," ll. 17-24. Among other similar poems there are "Verses," "To his Honour'd friend, Sir Thomas Heale," and "To the most accomplisht Gentleman Master Michael Oulsworth." Most of such poems are relatively brief and very effective epigrams.

'Tis but a flying minute,
    That I must stay,
    Or linger in it;
And then I must away. . . .

How many lye forgot
    In Vaults beneath?
    And piece-meale rot
Without a fame in death?

(5-8, 13-16)

Then come the stanzas given before, and on the page there rises, as his title proclaims, "His Poetrie his Pillar," his "living stone." Such acclimatizing of Horace to the earth under an English parish church sounds the true Herrick, and Jonsonian, tone. The same approximation of the Roman to the English will be found in the last poem but one in the *Hesperides*, "The pillar of Fame." But another poem, an epigram "On himselfe" says it all most clearly and is not the least attractive among the 1,130 poems of the *Hesperides*.

Live by thy Muse thou shalt; when others die
Leaving no Fame to long Posterity:
When Monarchies trans-shifted are, and gone;
Here shall endure thy vast Dominion.

The trans-shifting of monarchies was all too evident by 1648, when Herrick's poems were published. Although the crisis of the times was not wholly distinct from the arguments of time, it may be set aside for the moment for consideration of one or two last remedies of Time. One was what I shall again call the Anacreontic shrug. It involves no ignorance, and in fact the Anacreontic view is never far removed from the sight of death. But why worry? Cowley's poem on "The Grashopper" occupies an area between Stanley's and Love-

lace's poems in terms of accuracy and freedom of imitation.
Cowley could write (in a poem published in 1656):

> But when thou'st drunk, and danc'd, and sung
> Thy fill, the flowry Leaves among
> (*Voluptuous*, and *Wise* withal,
> *Epicurean Animal!*)
> Satiated with thy *Summer Feast*,
> Thou retir'st to endless *Rest*.[34]

It may well be that such an "Innocent Epicurean" is indeed
"Happier than the happiest *King*" (10). But better remedies
than summer sleep were required in the Cavalier winter.

In the end, we observe that the threat of time to the good
life was best opposed by reliance on the good life itself. Time
was transcended not by a centrifugal flight to an outer
sphere but by centripetal transcendence to the good man, to
integrity, and to the art of poetry as well as to that last art,
of being oneself. The centripetal movement is very much a
Horatian motif, as we can see from the conclusion of his ode
to Maecenas (III. xxix), speaking of defenses against fortune.

> What is't to me,
> Who never sail in her unfaithful Sea,
> If Storms arise, and Clouds grow black?
> If the Mast split and threaten wreck,

---

[34] Cowley, ll. 28-34. My favorite Anacreontic philosopher is the Wife
of Bath, that *"Epicurean animal,"* whose values one regrets needing to
dismiss.

> But, Lord Crist! whan that it remembreth me
> Upon my yowthe, and on my jolitee,
> It tikleth me aboute myn herte roote.
> Unto this day it dooth myn herte boote
> That I have had my world as in my tyme.
> But age, allas! that al wole envenyme,
> Hath me biraft my beautee and my pith.
> Lat go, farewel! the devel go therwith!
>   (D 469-476)

149

Then let the greedy Merchant fear
    For his ill gotten gain;
And pray to Gods that will not hear,
While the debating winds and billows bear
    His Wealth into the Main.
For me secure from Fortunes blows,
(Secure of what I cannot lose,)
    In my small Pinnace I can sail,
        Contemning all the blustring roar;
        And running with a merry gale,
  With friendly Stars my safety seek
Within some little winding Creek;
        And see the storm a shore.

                        (trans. Dryden, st. x)

The defense by art was to raise the indestructible monument. The defense by the good life is to curl into itself, *mediis tranquillus in undis*, or, in Horace's image here, from out of Fortune's sea.

Often the good life seems to preserve the happy life against time. Herrick sometimes appears to think that to "live merrily" answers Time's arguments, and Death's as well (see "Best to be merry"). But more was required than that, as the best poems by him and others show. And it would be hard to illustrate the requirement better than by returning to the subject of the country life. The country life was the retired life, the contemplative life. And each Cavalier poet at some stage, either seeks retirement or finds his values in the country. What is implied is retreat from the world and the times, from the court and the city, to oneself and to the country. The country poems provide us with the centripetal transcendence of Horace writ large. Marvell's *Garden* and *Upon Appleton House* at once conform to the tradition and break from it. We may begin, therefore, with consideration of the earlier

Royalist retirement if only because that of Fairfax and Marvell was to come later.

As is well known, the praise of country life in seventeenth-century poetry owes most to what is a distortion of Horace's second epode. In the original, the praise comes from Alfius, a usurer, who ends (67-70) by going back to his cynical trade after his long sentimental and hypocritical sigh for the country. In my view, it is a mark of Horace's sanity that he could allow such an undercutting of his own somewhat simplified treatment of life on the Sabine farm. In Cowley's delightful essays, we ultimately come to the same mature questioning of one's favorite values. Many poets before Cowley either dropped the ironic conclusion of *Epodes*, ii in translating it (as did Thomas Stanley) or simply ignore the irony of the situation. Yet others could best solve the problem by writing their own poems like *Epodes*, ii: up to the ironic conclusion. Other models could readily be found to illustrate the commonplace, *Rustica vita optima*,[35] and after all, conditions in classical Rome or malarial London in the seventeenth century really were not salubrious. Among the handsomest of the many country poems, Herrick's *Country Life: To his Brother, Master Thomas Herrick* is probably one of his most neglected pieces. The time of writing falls at some period before the Cavalier summer ended, at a time when even loyal Cavaliers would speak of "that too-true-Report, / Vice rules the Most, or All at Court" (89-90). Nowhere does Herrick show in himself more of Ben Jonson than in this poem. For if we miss the splendid conjuring of "To Penshurst," we discover the

---

[35] *"Rustica vita optima,"* cap. ciii, *Synopsis Communium Locorum*, 3rd ed. (London, 1719; 1st English ed. 1700). The works referred to are the standard ones: Horace, *Epodes*, ii; Virgil, *Georgics*, ii. 458-89; *Culex*, 58-97; Seneca, *Hippolytus*, ll. 483-539; Tibullus, i. x. 39-42 (also much in the whole elegy, formerly numbered xi); and Martial, i. lv.

centripetal movement to the inner resources of the good man. His brother may count his blessings.

> But thou at home, blest with securest ease,
>> Sitt'st, and beleev'st that there be seas,
> And watrie dangers; while thy whiter hap,
>> But sees these things within thy Map. (69-72)

Retreat and the country, almost in itself, protect his brother. But without the *integer vitae*, without the *strength* to move within, the centripetal movement would be impossible. Herrick's brother possesses a wide-eyed and indeed a dry-eyed fortitude.

> But thou liv'st fearlesse; and thy face ne'r shewes
>> Fortune when she comes, or goes.
> But with thy equall thoughts, prepar'd dost stand,
>> To take her by the either hand:
> Nor car'st which comes first, the foule or faire;
>> A *wise man ev'ry way lies square.*
> And like a surly *Oke* with storms preplext;
>> Growes still stronger, strongly vext.
> Be so, bold spirit; Stand Center-like, unmov'd . . .
>> (93-101)

Only a person fully read in Cavalier poetry could possibly guess that those lines are Herrick's rather than Jonson's. The same moral fiber will be found, although less rigorous concentration of language and tautness of syntax. Herrick shows less of the Church Militant, more of settling the good life in terms of moderation, that golden mean ("auream . . . mediocritatem") of Horace, *Odes,* II. x. Or as Herrick himself put it,

> This Theame,
> To shun the first, and last extreame.

Ordaining that thy small stock find no breach,
  Or to exceed thy Tether's reach:
And to live round, and close, and wisely true
  To thine owne selfe; and knowne to few.

(131-36)

In yet another poem, "On himself" ("A Wearied Pilgrim")
we hear again, as not seldom, that Jonsonian note.

He lives, who lives to virtue: men who cast
Their ends for Pleasure, do not live, but last.

(9-10)

When the self-reliance of the *integer vitae* shades off into
received concepts of virtue, the ethical stance may be as easily
termed Christian as anything else. When the Cavaliers approached the good life in such terms, they entered the common waters of seventeenth-century poetry. But they mostly
kept to their own preserves, as I think a comparison of three
elegiac poems would show: Donne's *First Anniversary*, Jonson's pindaric ode on *Sir Lucius Cary, and Sir H. Morison*,
and Milton's *Lycidas*. In Donne, there is a certain hysteria
but also a certain ecstasy. The fear is not so much of time as
of its direction toward imminent disaster. The solution, if it
can be found, would be to move to the higher realm of the
"blessed maid" whose death encourages such bleak medita-
tions. *Lycidas* shows Milton worried over time and his own
"unripe" state, over his elusive pursuit of fame, and over the
sorry behavior of the pastors of the time. But he can welcome
the next day and fresh scenes, once he has seen off Lycidas to
heaven in that rapt vision of the marriage feast of the Lamb
of God. Milton may return safely to time *after* having gone
clean out of it. To Jonson, and to his Sons, it was necessary
first to turn in, to transcend the times and defeat Time by
moving into one's secure virtue. Even the strange, or maca-

153

bre, simile of the child returning to the womb at the beginning of Jonson's ode makes the same point. The reduction to essentials is a kind of moral alchemy, producing an elixir of life. And finally, what I have perhaps not emphasized enough, after the movement inward, it is possible to return (in the fashion opposite to Milton) to the times, to Time, or even to eternity. To quote Jonson's passage yet again,

> Accept this garland, plant it on thy head,
> And thinke, nay know, thy *Morison's* not dead.
> Hee leap'd the present age,
> Possest with holy rage,
> To see that bright eternall Day:
> Of which we *Priests*, and *Poëts* say
> Such truths, as we expect for happy men.   (77-83)

But such movement outward is only possible after turning inward to "small proportions" and "short measures." Once we possess the "flowre of light," we are capable of seeing "that bright eternall Day." Once the transcendence inward is complete, we may go out speaking the "truths" that may be expected "for happy men"—here a *vita beata* in a religious sense. That kind of happy manhood can be earned only after turning inward to the good manhood that Jonson takes such pains to create in his poetry.

The remedies of time rest ultimately on the virtues of the good life, not on the enjoyments of the happy life. Lasting enjoyment is premised on (and considerably pruned by) a moderation so close to the *integer vitae*, to a self-sufficiency, or to a Christian morality that the heart of Cavalier poetry is the heart of the good man. Sooner or later, regulation of the self and return to the regulated self are discovered to be the means for coping convincingly with time. The old commonplace that to rule oneself is the greatest dominion meant more to the Cavaliers than to any other poets of the century.[36] Donne's policy of exploring within the private world for what may belong to larger worlds was not followed by Jonson and his Sons. They explored the social world in order to discover their own inner resources. At times it seemed enough to confirm that the resources existed. At other times, it might be necessary to "leap the present age" for eternity. Although the Cavalier conception of life will hardly be thought to exclude all others, it claims our admiration and our poetic faith. Indeed, in one sense the Cavaliers wrote better than they knew during the first three decades of the century. Time would become a problem by bringing them sorry times. One doubts that mankind has ever known an era of halcyon calm, but one can be certain that as they approached mid-century the Cavaliers knew their vessels had entered stormy seas, and their state to be buffeted by heavy northern winds. They had entered the winter of their discontent.

[36] *"Sibi imperare, imperium maximum,"* *Synopsis Communium Locorum,* cap. cxviii. The first lines quoted are cited, "Ex Auson. (si memini)," but one cannot say much for the compiler's memory, or his accuracy. The other places he cites are (in corrected form): Horace, *Odes,* ii. ii. 9-10; *Satires,* ii. vii. 83-88; Claudian, *On the Fourth Consulship of Honorius,* ll. 255-68; Seneca, *Thyestes,* ll. 345-68, 388-90, and 470; Martial, ii. liii; Propertius, ii. xxiii. 23-24; Ovid, *Amores,* ii. xii, 5; *Remedia Amoris,* l. 54. The purpose of supplying such citations in this chapter (and others) has been partly to spare other students the time needed to run these things down, and partly to define the nature of Cavalier "classicism" as opposed to that of a Donne, Milton, or Dryden.

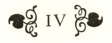

IV

ORDER AND DISORDER

A sweet disorder.
—Herrick

Thou seem'st a Rose-bud born in Snow,
A flowre of purpose sprung to bow
To headless tempests, and the rage
Of an Incensed, stormie Age.
—Vaughan

THE SEARCH FOR informing principles of seventeenth-century literature has led critics to a number of polarities and illuminating accommodations. To the men and women alive between the sixteenth and the eighteenth centuries, such polarities as God and man, sin and grace, art and nature, or reason and passion very clearly occupied a centrality to their thought that most literary critics today have not found it necessary to consider so crucial. The conceptions most fruitfully employed in our time, and perhaps also overdone in our time, have been appearance and reality, originality and tradition, and that harmony which is *discordia concors* or *concordia discors*.[1] All these conceptions have proved very

---

[1] A small pedantic dispute turns on which of the two Latin phrases is the correct. They both are. Horace speaks of *concordia discors* in *Epistles,* I. xii. 19; Manilius of *discordia concors* in *Astronomicon,* I. 142 (which Thomas Creech translated "agreeing disagreement"); and Ovid, *Metamorphoses,* I. 433, of *discors concordia*. But other formulations will be found, as in Lactantius, *Institutiones Divinae,* II. ix. 17: "poetae discordia concordia mundum constare dixerunt" ("poets say that the world is established on discord-concord"); or Seneca, *Epistulae,*

useful, as also along with them certain others, for example the one and the many, which are variously said to be neo-Platonic, syncretistic, or "spiritual." In choosing to introduce (as the title of the chapter shows) another pair of words, my aim is only to return to the experience treated by the poets and, as far as possible, to use their terms to describe that experience. By quoting a phrase from Herrick, I hope that most of my readers will have recalled with pleasure his "Delight in Disorder." And some may be led by Vaughan to remember that in the induction to *Every Man Out of His Humour* Asper speaks for the poet in wishing to expose "the ragged follies of the time." Examination of "order" and "disorder" in the *Oxford English Dictionary*, which makes us all so wise, will show the many senses—ecclesiastical, moral, military, abstract, and figurative—in which these words operated during the seventeenth century.

i. "Pure Enough, though Not Precise"

The problem with my choice of terms does not fall so much in the need to demonstrate their validity as to keep them under control. So let us try to adapt "order and disorder" as quickly as possible to the terms of the preceding chapters. Aristocratic mores by definition imply a social class and its behavior, and only a person unacquainted with aristocrats (or human nature) would think that they do not prize disorder within the orderly social frame. Perhaps Jonson and Lovelace set some of the terms of order most firmly, whereas Suckling and Rochester admit the most disorder. Further, the good life (*vita bona*) tends to be an orderly life,

---

LXXXIV. 10: "concentus ex dissonis." Quintilian in his *Institutes* (I. x. 12) sums up a major source and the end to which phrases drive: "illa . . . dissimilium concordia, quam vocant ἁρμονίαν" ("that concord of discordant elements which they [the Pythagoreans] call 'harmony' ").

certainly a life of interior order, integrity, whereas the happy life (*vita beata*) borrowed more from variety and exceptions. And the more the happy life came under threat, the more it was likely to be rescued from our foe Time by recourse to forms of order or transcendence in the good life.

Well and good. But we quickly discover that there often comes a denial in the value of order when it is considered excessive or even unnatural. And that "when" admits some pretty wide limits. Jonson's song from *Epicoene* (i. i), "Still to be neat, still to be drest," may or may not express his deeply held values, but it certainly embodies a motif of ancient lineage.[2]

> Though arts hid causes are not found,
> All is not sweet, all is not sound.   (5-6)

What one desires in a woman's appearance turns out to be rather that which "makes simplicitie a grace":

> Robes loosely flowing, haire as free:
> Such sweet neglect more taketh me,
> Then all th' adulteries of art.
> They strike mine eyes, but not my heart.
>
>   (9-12)

As any one can see from its formal properties, however, the poem does not strictly embody such "sweet neglect." The pattern of accent that may be marked in the first, third, and last lines of each of the two stanzas surely exceeds what would be required for the opposite argument.

$$' x \mid x \;' \mid \mid \;' x \mid x \;'$$
$$— \mid — \mid \mid x \;' \mid x \;'$$
$$' x \mid x \;' \mid \mid — \mid x \;'$$

[2] See Kirby Flower Smith, "On the Source of Ben Jonson's Song, 'Still to be Neat,'" *American Journal of Philology*, xxix (1908), 133-55.

Jonson's art "adulterates" the simple plea for nature, and as so often in the pleas for disorder by these poets, an underlying order justifies what seems a license for confusion.

To continue with the familiar, we may turn to Herrick's version of the subject, "Delight in Disorder." He tells how the "sweet disorder" and the "wilde civility" of the clothes of a very lively young lady (whose stormy legs lead to a "tempestuous petticote")

> Doe more bewitch me, then when Art
> Is too precise in every part. (13-14)

That click of accent wonderfully conveys the precision it detests. And yet the formal properties of the poem again reprieve the delighted Robert Herrick from any simple charge of disorderly conduct. Those ribbons flowing "confusedly" represent no principle of artistic or social anarchy. The poem makes use of that old sixteenth-century friend of ours, the blazon, the itemizing of a lady's attractive features from top to bottom. And adding to this order of the clothing (each item suggesting the lady's physical and moral presence, and its effect on the male observer), the blazon is framed by the initial and closing couplets, which alone make up a predication. We may most temperately say that a "sweet disorder" regulated one kind of nature by one kind of art, and that a "wilde civility" freed one kind of art by one kind of nature. Herrick is after a kind of moderation, or balance.

His search emerges more clearly in a less familiar (and less successful poem on the whole), "What kind of Mistresse he would have," an old recipe for daydreaming poets. She should be

> Pure enough, though not Precise:
> Be she shewing in her dresse,
> Like a civill Wilderness;

159

That the curious may detect
Order in a sweet neglect.   (4-8)

We seem almost to encounter a *discordia concors* of moral
Lucretia and loose Thais (which is *almost* what Herrick says
in lines 15-16). But Herrick's mean, if not golden, is sterling,
and it emerges in the passage quoted in the special poise of
"Pure" and "Precise" or the qualifications, "civill Wilder-
ness" and "sweet neglect." And he concludes,

Be she such, as neither will
Famish me, nor over-fill.   (17-18)

Perhaps one can say a very few less obvious things about
poems such as these—or "The Argument" prefixed to *Hesperides*. One of them may easily be represented by Herrick's
line, "Pure enough, though not Precise." The point is that,
however *pure* his mistress will be, she will *not* be a Puritan.
The word "precise" is a critical word in both of Herrick's
poems quoted, and once again the *Oxford English Dictionary*
shows the ecclesiastical and colloquial applications of "precise" and "precisian" in the late sixteenth and seventeenth
centuries. They are, adjective and noun, synonymous with
"puritan." Herrick clearly has in mind as a positive code a
Cavalier style of dress, and life, so distinguishing what he be-
lieves in from the style of dress and life of the Precisians. The
very "Argument" to the *Hesperides* provides a declaration of
intent to be Anglican and Cavalier: "I sing of *May-poles*,
Hock-carts, *Wassails*, *Wakes* . . ." (4). About the time he
wrote, or at least published, that argument, there were those
so far from singing of "The Court of *Mab*, and of the *Faerie-
King*" (12) that they had imprisoned Charles I and would
shortly behead him. The *Hesperides* came into the world between the Parliamentary ordinance against such festivals as
Christmas and the unprecedented judicial execution of an
English king.

Although Butler was to speak of the Civil Wars as a time when "men fell out they knew not why," and although words like "Cavalier" and "Puritan," invented to insult one's enemies, may prove misleading, still one group fought another in the seventeenth century. What may be harder for some readers to accept is that Herrick's poems are involved in some way in the not-so-sweet disorders of the time. We must acknowledge that often the *style* of doing something brings greater shock than change concealed by adherence to old forms. Now until about 1648 or 1649, there was furious debate in England over such matters as what we would consider the furnishing of the church: where the communion table should stand, whether there should be a rail, what the table should be covered with, and what the minister should wear while standing in front or behind it. How to dress? That was indeed a furiously debated question. How much "disorder" in church dress was "sweet"? The common comparison among Anglicans was of the Church of Rome to the Babylonian whore decked in false finery, and of the Church of Calvin to "country drudges," as Donne put it in *Satyre III*: "plaine, simple, sullen, yong, / Contemptuous, yet unhansome" (54, 51-52). The style of one's mistress' dress very much involved the dress of "our Mistresse faire Religion" (5), as Donne recognized. And so Herrick. He wants his mistress "Pure enough" but not Puritan, no Precisian. The evidence of Herrick's popularity has been interpreted variously, but it is at least a fact that the *Hesperides* was published twice in 1648 —something extraordinary for a collection of anything other than divine poetry in the century—but then not again until the nineteenth century. Why the sudden popularity and then the silence? In fact, there was not silence but repeated whispers. *Witt's Recreations* included above sixty of his poems in editions of 1654, 1663, and 1667, and although the bibliographical situation is confusing, other editions of these *Recre-*

161

ations appear to have been published. Moreover, "between 1649 and 1674, almost a hundred of [his poems] were reprinted, usually without credit to him, in 28 different collections."[3] My inference is that he was popular among Royalists and rejected by "Precisians." Poems quoted at the end of the last chapter showed Herrick to exercise a Jonsonian seriousness. And what I argue here is a timeliness upon appearance.

At this juncture we should recall that Jonson's song rejects "all the adulteries of art," of a woman *always* ("still") "pou'-dred, still perfum'd." It is impossible to believe that Jonson imagines a woman other than a lady of the court or a woman of the town, in either of the two possible senses. All one need do to understand so much is to compare this "Lady" with the fine whore attacked in "An Epigram on the Court Pucell." The adulteries of art may imply more than cosmetics. And yet criticizing adulteries of art does not make Jonson a Puritan any more than Herrick was a Catholic for preferring fancy dress. A degree of paradox may seem to enter into such matters. Herrick can write, "Art presupposes Nature" ("Upon Man," l. 3), an unexceptionable sentiment. Yet he can also write "Art above Nature, to Julia."

> Next, when those Lawnie Filmes I see
> Play with a wild civility:
> And all those airie silks to flow,
> Alluring me, and tempting so:
> I must confesse, mine eye and heart
> Dotes less on Nature, then on Art.
>
> (13-18)

The echo of Jonson's song is stronger here than in "Delight in Disorder." And yet the conception is contrary to Jonson's explicit statements and it employs a differing adjustment of

---

[3] J. Max Patrick, ed., *The Complete Poetry of Robert Herrick* (New York, 1968), p. xii.

presumptions from that of "Delight in Disorder." And yet again it must be said that although the cast of the ideal has changed, the woman is still no Precisian. Similarly, nothing in the Cavalier vision of life required that poets think the court or even a king free from evil. Beaumont and Fletcher's tragicomedy, *The Maid's Tragedy*, like Suckling's *Aglaura*, is no less Cavalier for raising fundamental questions about kings and kingship. I very much believe that for their ideas, their poems, and their lives, the Cavaliers were the better off for not being "too precise in every part."

Our poets not unnaturally believed in principles whose bounds needed to be set. Mere disorder was chaos, a threat to Nature herself. Art was necessary to civilize the wilds of Man and Nature. (Only Cotton could appreciate the bleak scenery near his estate sufficiently to write an English poem on the area, *The Wonders of the Peak*.) And yet order alone was sterile, requiring a degree of variety, disorder, to assure life and movement. Once such an ambiguously evaluated polarity as order and disorder is set, the questions thicken. What is the degree of admissible variety commensurate with order? What kind of order is natural? What are the limits of an ordering art before artifice stifles? And so on. I submit that such questions exercised the whole seventeenth century and have lost none of their pertinence with the passage of time, although the terms and assumptions may change. The Royalists and the Parliamentarians and the army each supported forms of order, each had concepts of liberty (that is, of stipulated rights) that seemed disorderly to the others, and each was willing to admit certain kinds of disorder in payment for an order it more highly prized. What could not be agreed upon was the kind or *locus* of social order that was desirable, or the proper liberty in church and state. Little can be found by a twentieth-century liberal to commend in any of the practical polities, all of them generally repressive, in the seventeenth century. But the Cavalier poets came upon truths that neither other poets nor the shifting rulers of the century understood enough to make practice feasible: an element of disorder was necessary to keep the larger order sound.

I return for almost the last time to the frontispiece and titlepage of the *Hesperides*. Set that great crown beside contemptuous references at the time to Charles Stuart. Consider on the frontispiece that Pegasus with cupid-like figures that scatter flowers or dance in a ring, and consider it against the debates in Parliament or the letters of Cromwell. Look at that poet fat like Horace. Look at those curls. Here is no closely cropped Roundhead, here no canting Precisian.[4] Here is a Cavalier concerned with time and mortality, believing his poetry will triumph over such larger forces as well as over the Precisians and Parliamentarians. Beneath the bust of the poet there will be found an eight-line Latin poem. They tell Herrick justly that he has excluded arms from his poetry ("excludis Versibus Arma tuis"). Herrick's vocations were rather those of the Church and Parnassus. But we can find particular significance in the word beginning the first and third verses. It suggests the relation of Herrick's poems to the age and suggests the major problem faced by Cavalier poets in *ordering* their lives: the problem is not now "Tempus edax rerum," Time the Devourer of things, but *tempora*, the times.

ii. "These Anxious and Dangerous Times"

No doubt the times are always bad. Things were always better than they are, and it is not only Cicero who worried

---

[4] Many of Herrick's admirers have not wished to think that bust to be of Herrick; it is of too fat a man. It may be another person represented, but Herrick's poetry does not suggest that he was costive and thin. After seeing the effigy of Lucius Cary, Viscount Falkland, in the

over "these anxious and dangerous times."[5] But when the en-during human lament enters poetry, we must, if we readers aspire to be Jonson's "understanders," consider how it enters and what role it plays. I wish to begin with a simplified prop-osition that will later require adjustment. As time worked most immediately and unfavorably on the happy life, so the immediate threat to the good life became the times. Let us summon again the spirit of the younger Democritus.

> To see wise men degraded, fools preferred, one govern towns and cities, and yet a silly woman over-rules him at home; command a Province, and yet his own servants or children prescribe laws to him, as *Themistocles'* son did in *Greece*; *What I will* (said he) *my mother wills, and what my mother wills, my father doth.* To see horses ride in a coach, men draw it; dogs devour their masters; towers build masons; children rule; old men go to school; women wear the breeches, sheep demolish towns, devour men, &c., and in a word, the world turned upside downward! *O viveret Democritus!*[6]

Now it may be thought that Burton is putting this on, that here we find merely the old tropes for disorder, lacking only birds in the water and fish in the trees. But in fact Burton possesses a genuine vision of his own day. He undertakes to show "that Kingdoms, Provinces, families, were melancholy as well as private men" (1, 79). He follows a paean on the

---

Church at Burford, I am resigned to believing that likable Cavaliers were stout.

[5] Cicero on "dubiis formidolosisque temporibus," *Verrine Orations,* II. v. 1. 1. In *Epistolae Ho-Elianae,* 11th edition (London, 1754) James Howell remarked that "The Times are so ticklish" and much else to that end (p. 291: 1. vi. 53); adding, "and you know, as well as I, what Danger I may incur."

[6] Robert Burton, *The Anatomy of Melancholy,* ed. A. R. Shilletto, 3 vols. (London, 1920), 1, 73.

fame of England with talk of "some bad weeds and enormi-
ties"; in fact, one wonders how Burton's "Satyricall Preface"
was received in the 1652 edition? He spoke of "a wise,
learned, religious King, another *Numa*, a second *Augustus*
[executed in 1649], . . . most worthy senators [Cromwell was
struggling with the Rump], a learned clergy [now seques-
tered], an obedient commonalty [Col. Sexby and others were
plotting against Cromwell's very life], &c." (1, 97). It is not
our century that first learned to add political two's and two's
or to distinguish the colored spectrum of events in the day.

Sir John Denham's *Cooper's Hill*, a poem of a degree of
greatness that none of his other original verse permits us to
expect, exemplifies the serious poetic concern with the times.
His first object of description is the City, as surveyed by the
King and rejected by the poet.

> Under his proud survey the City lies,
> And like a mist beneath a hill doth rise;
> Whose state and wealth the business and the crowd,
> Seems at this distance but a darker cloud:
> And is to him who rightly things esteems,
> No other in effect than what it seems:
> Where, with like hast, though several ways, they run
> Some to undo, and some to be undone;
> While luxury, and wealth, like war and peace,
> Are each the others ruine, and increase;
> As Rivers lost in Seas some secret vein
> Thence reconveighs, there to be lost again.
> Oh happiness of sweet retir'd content!
> To be at once secure, and innocent.[7]

[7] Lines 25-38 of the 1668 version of the poem. In *Expans'd Hieroglyphicks* (Berkeley and Los Angeles, 1969), Brendan O Hehir has very
carefully, one might say finally, distinguished the different versions: my
quotation comes from his "B Text," his "Draft IV," pp. 137 ff. I have
been reminded that in the first edition, Denham described London yet

The verses of course describe a place, the City of London, rather than the times. But Denham develops a contrast between the present city and the past court at Windsor, even while preferring a country ideal.

Many kinds of evidence in *Cooper's Hill* testify to Denham's concern with the times. In particular, there is the well-known allegory of the hunting of the stag for the death of Strafford (235-322). To adapt Marvell's words, Denham seems to think Strafford too good a man to be fought for and to wish to leave the whole business to God and the King. Less obvious, or less familiar, details also convey the character of the times. The poem concludes with the Horatian figure for the state in turmoil, a river in flood.

> When a calm River rais'd with sudden rains,
> Or Snows dissolv'd oreflows th'adjoyning Plains,
> The Husbandmen with high-rais'd banks secure
> Their greedy hopes, and this he can endure.
> But if with Bays and Dams they strive to force
> His channel to a new, or narrow course;
> No longer then within his banks he dwells,
> First to a Torrent, then a Deluge swells:
> Stronger, and fiercer by restraint he roars,
> And knows no bound, but makes his power his shores.
>
> (349-58)

It is a very pessimistic ending. We have been well taught in recent years to read Denham's masterpiece as an essay on harmony proceeding from discords.[8] This reading certainly

---

more critically: "I see the City in a thicker cloud / Of businesse, then of smoake; where men like Ants / Toyle . . ." (28-30).

[8] See Earl R. Wasserman, *The Subtler Language* (Baltimore, 1959), pp. 35-88. Brendan O Hehir's life of Denham, *Harmony from Discords* (Berkeley and Los Angeles, 1968), extends Wasserman's seminal study, especially emphasizing the political character of the poem. See also O Hehir's book mentioned in n. 7, above.

accounts for most of the poem, and it is allowable for some earlier versions of the ending, in which the possibility is raised of reconciliation between the King's party and his opponents. Like many a Cavalier (and Puritan), Denham simply wished to view from afar in the early stages of struggle and conflict. But by 1655 his attitude had shifted enough to make him regard the Civil Wars and the loss of kingly rule as a disaster of state for which the flooding river is an old trope and a popular image in contemporary writing. The poem retains, however, that very Cavalier sense of the vices as well as the virtues of princes, and so also the Horatian impulse to seek the golden mean.[9]

How Cavalier Denham's emphasis is may be understood by comparing his poems with some of those by Donne and Dryden. Donne was not happy with the Court, or rather his *Satyres* castigate the court; but his lyrics do not show that the country afforded him pleasure, either. He wishes for a fine and private place in the city, to go to a town house, not to the "happiness of sweet retir'd content." The Cavalier vision entailed either staying on the scene and accepting the courtly, aristocratic constitution, without necessarily affirming its perfect health; or it entailed going to the country, as we have repeatedly seen. In this the Denhams also differ from the Drydens. Dryden seems to have thought the country a fine place to fish, but not a good place to live. His only known lengthy absence from London after the Restoration was caused by the Great Plague. He wrote no country poem quite like *To Penshurst* or even anything as directly descriptive as Waller's *Upon St. James's Park*. Of course no seventeenth-century poet, Denham and Cotton excepted, goes to

---

[9] I have dwelt enough in earlier chapters on these images as used by Horace; for other seventeenth-century usages, see O Hehir's notes to this passage and the annotations on *Threnodia Augustalis* in *The Works of John Dryden*, III, 302-303, and on the passages there referred to.

lengths of description like that we might hope to find in Thomson or Cowper. And Jonson does not put every stone of the Sidney estate into place. But as I write, with a picture on the wall before me of the estate of John Driden at Chesterton, and as I look at Dryden's great poem, *To My Honour'd Kinsman*, I can but conclude that only an imagination of far greater vivacity than mine could raise the picture from the poem. Denham, on the other hand, shows the Cavalier tendency to observe the country, once one has gone to it. We recall some of Marvell's descriptive images, and we shall see that Vaughan could look at as well as see into things in his Cavalier poems. Descended from Horace, the "Serene Contemplator" of many Cavalier poems is the grandfather of that Spectator of the country in the eighteenth century.

iii. Jonson's Definition of Virtue, "Time, and Chance" and Afterwards

The treatment of the times, like the treatment of almost everything in Jonson's poetry, involves fundamental moral rather than political issues in the usual sense. (The profound political implications of his ethical drive must be left for consideration on another occasion.) Jonson dismisses such Puritans as Tribulation Wholesome or Zeal-of-the-Land-Busy for their hypocrisy and greed. They are not depicted as those troublesome Saints Militant who finally routed the King and his Cavaliers. Because the issue of the times seemed moral to Jonson, and because he responded with such remarkable moral vigor himself, his poems seem particularly unpartisan. He is fully immersed in society. He fully needs to be there— how much information did he have for Drummond about events outside London and the great estates? But being Jonson, he also needs in particular to be himself, to assert that formidable combination of egotism and integrity against the

169

world. Sometimes his sense of the world's pravity, or of erring men and women of his time, expresses itself poetically in epigrams that seem about to emerge as "characters," whether satiric as "On Don Surly" or panegyric as "To Robert Earle of Salisburie" ("Who can consider"). But we always sense a connection with the times. We may be uncertain who the "Court Pucell" is, and in fact I believe her a composite, but a composite of women observed at the time, not a timeless (in more than one sense) female type like Webster's Fair and Happy Milkmaid. Often Jonson's epigrams seem miniature scenes, germs of some play he was about to write on the times, as for example "On Giles and Jone," with its highly amusing version of the *concordia discors* motif; or "On Mill. My Ladies Woman," a progress of venery to marriage. In such poems the moral judgment on people, acts, and times retains the curious blend of sportiveness and rigor that marks Jonson's lighter moments.

His epistles differ from the epigrams in that the moral tone is more earnest and the satiric duel with the times is totally committed, serious to the point of death. Perhaps we can begin to observe what may be termed the ethical decorum of Jonson's poetry: viewing different men and different times differently produces a range of varying ethical responses appropriate to one condition but not another. Here is Jonson on the struggle between virtue and the present times in his *Epistle to Katherine, Lady Aubigny.*

> 'Tis growne almost a danger to speake true
>   Of any good minde, now: There are so few.
> The bad, by number, are so fortified,
>   As what th'have lost t'expect, they dare deride.
>
> (1-4)

The danger is clear. So is Jonson's dismissal of the bad in the class of the person he addresses, that is great wives.

Let who will follow fashions, and attyres,
　　Maintayne their liedgers forth, for forraine wyres,
Melt downe their husbands land, to poure away
　　On the close groome, and page, on new-yeeres day,
And almost, all dayes after, while they live.　　(71-75)

Such a vision of the court and town sends the Satirist in
Donne's *Satyres* reeling home in nausea. Not so Jonson. His
sword will not sleep in his hand. And he knows that the virtue
of the good man arms him with a real, active, indeed a militant energy that "can time, and chance defeat." His praise of
the lady obviously turns on what he values most.

　　Wherein, how more then much
Are you engaged to your happy fate,
　　For such a lot! that mixt you with a state
Of so great title, birth, but vertue most,
　　Without which, all the rest were sounds, or lost.
'Tis onely that can time, and chance defeat:
　　For he, that once is good, is ever great.　　(46-52)

Three things make the passage a microcosm of Jonson's view
of himself in his times. The first entails the danger from
"time, and chance." The second appears in the combativeness
of that "vertue." And the third will be found in that quin-
tessentially Jonsonian aphorism concluding the passage: that
still integrity which triumphs unmoved (*immota triumphans*
is the commonplace) over its adversaries.

　　Jonson's ethical decorum requires a different tone when his
attention turns to a view of serious corruption. There will
not be found much room for the man of integrity in the
world of *An Epistle to a Friend, to Perswade Him to the
Warres.* One's own country has certainly fallen on evil times
when the only remedy for a young man becomes resolute
fighting with the expectation of death in the religous wars of

the Continent. Jonson begins with ten lines rousing his friend to vital, honorable action, and then for 130 lines (11-140) he sets forth a series of satiric prospects, each beginning, "Looke, . . ." or "See, . . ." or "Who can behold . . ." and finally, "I'le bid thee looke no more, but flee, flee, friend, . . ." (129). Jonson follows Burton in depicting a social disorder in which roles are reversed. But what is so remarkably Jonsonian is that his depiction drives irresistibly to ethical ends.

> Our Delicacies are growne capitall,
> And even our sports are dangers! what we call
> Friendship is now mask'd Hatred! Justice fled,
> And shamefastness together! All lawes dead
> That kept man living! Pleasures only sought!
> Honour and honestie, as poore things thought
> As they are made! Pride, and stiffe Clownage mixt
> To make up Greatnesse!   (37-44)

That this form of moral poetry involves what I have called an ethical decorum (some might say simply an art) can be shown, I think, from that passage in which Jonson develops the satiric commonplace that it is difficult *not* to write satire in such times.[10]

> O, these so ignorant Monsters! light, as proud,
> Who can behold their Manners, and not clowd-
> Like upon them lighten? If nature could
> Not make a verse; Anger; or laughter would.
>
>   (59-62)

This passage is from the last section but one (141-74) of the poem, in which the poet summarizes and comments on the disorder of vices he has been portraying in his satiric prospects for his friend. No small poetic part of the effectiveness

---

[10] Cf. Juvenal, *Satires*, 1. 30, "difficile est saturam non scribere," the *locus classicus*.

of these prospects derives from that very Jonsonian movement from metaphor to image, using an imaginative burst to
"lighten" on his subject:

> or have we got
> In this, and like, an itch of Vanitie,
> That scratching now's our best Felicitie?
>
> (142-44)

Such felicities (if Father Ben will permit the term) of
style provide local features of the total conception of the
poem, but they are less immediately felt than the sense of
integrity in the poet and his friend. Speaking of a man who
is at once a sycophant and a pander, Jonson compresses much
of the poem into half a dozen lines.

> As a poore single flatterer, without Baud
> Is nothing, such scarce meat and drinke he'le give,
> But he that's both, and slave to both, shall live,
> And be belov'd, while the Whores last. O times,
> Friend flie from hence; and let these kindled rimes
> Light thee from hell on earth.   (158-63)

The Ciceronian lament over the times and men's behavior
(*O tempora! O mores!*) rings out clearly: but what is the
solution? Jonson's first step in positive advice is the important one. He seeks to confirm the virtue he has found in his
friend.

> Goe quit 'hem all. And take along with thee,
> Thy true friends wishes, *Colby* . . .
> That thou dost all things more for truth, then glory,
> And never but for doing wrong be sory;
> That by commanding first thy selfe, thou mak'st
> Thy person fit for any charge thou tak'st;
> That fortune never make thee to complaine,

173

But what she gives, thou dar'st give her again;
That whatsoever face thy fate puts on,
Thou shrinke or start not; but be always one.

(175-86)

It is remarkable how the style alters in this passage. The earlier rough, explosive, compacted, much-stopped style—in fact, the strong line—has yielded to a no doubt still forceful, but a fluent, end-stopped measure that would have served Jonson in one of his plays. Jonson's ethical decorum of harsh lines for satire and smoother for affirmation sometimes involes a music heard only by the practiced ear, but it is a music that a little study teaches. And we see, moreover, even in this lengthy satiric epistle, that Horatian movement from without to within, now from other men to the particularly Jonsonian "be always one."

The resolution of the poem presents something of a problem. We can recognize easily enough that the *integer vitae* suffers threat from such "times" and from the buffets of "fortune" and "fate." But the voice of the poet urges the friend to flee while the poet stays. It almost seems that Jonson has confidence in his own staying powers, and of course he is no longer young enough to return to the field. But the friend is presented a course different from staying.

These [verses] take, and now goe seeke thy peace in Warre,
Who falls for love of God, shall rise a Starre.   (195-96)

The friend's course is to be that of Sir Henry Morison of the pindaric ode. Moreover, like Morison and like Shakespeare, the friend's victory entails the kind of metamorphosis spoken of in the last chapter, with once again something of the classical idea of apotheosis working with something of the Christian glorification. The good death proves the good life. The ordered heart will triumph over the disordered world: "be always one."

174

In the poems we have been examining, Jonson's conception of the "follies of the time" turns on a fundamental contrast between an order within the good man and a teeming disorder without. He has paid little attention to specific events in their historical aspect. He appears to have wished to draw all his enemies into one army and to take on the times single-handedly without much regard to the individual captains or the actual terrain of the times. With the joining of Royalist and Parliamentary forces, however, and subsequently with the execution of the King and all that followed, events in time began to assume a greater historical character. Puritans led the way by seeing in contemporary events the writing of God's finger.[11] The discovery that the times one lives in are as much historical as the times of Moses, Pericles, or Augustus must be attributed to the period of the Civil Wars and was chiefly, I think, a lesson taught the nation by millennially minded Puritans.[12] But quite apart from the discovery's being taken up by Royalists like Cleveland, it must be said that Herrick expressed in his poetry a clearer and more explicit historical view than Jonson did. Jonson's view of the times was essentially moral. He could have walked into the house of Juvenal or have taken up the mantle of Jeremiah without breaking a stride. With Herrick, we begin to see that the times are the present, historical moment in church and state.

One of the first things we encounter also contrasts with Jonson: division of mind. Jonson saw as it were the enemy coming, stood, and fought. He advised his friend to flee a

[11] The term "Puritan" refers in my usage very generally to those Precisians who opposed the King and the episcopacy; it is a term fundamentally vague, altogether unsatisfactory, and wholly indispensable.

[12] On the Puritan sense of history and the times, see two excellent books, Arthur Barker, *Milton and the Puritan Dilemma* (Toronto, 1942); and Michael Fixler, *Milton and the Kingdoms of God* (London, 1964). For a strictly historical study of the Royalists, see Paul H. Hardacre, *The Royalists During the Puritan Revolution* (The Hague, 1956).

175

country so bad that a mortal battle was to be preferred. Later writers are simply unsure what course they have, and in this respect Herrick first takes that path to dubiety where Andrew Marvell claimed final possession. As Herrick puts it in his verses "Upon the troublesome times,"

> Where shall I goe,
> Or whither run
>   to shun
> This publique overthrow?   (5-8)

In a few years Cowley would ask himself the same thing, when a contemplative life in the English countryside turned out to be quite different from what he had expected. Bunyan also asks where man may fly, what he can do to be saved. The question haunts Restoration drama. The dilemma accompanied Rochester into debauchery, nihilism, and a death-bed conversion. Sometimes one lapsed into a state of numb resignation, with perhaps a catch at hope, as in Herrick's poem, "The bad season makes the Poet sad."

> Dull to my selfe, and almost dead to these
> My many fresh and fragrant Mistresses:
> Lost to all Musick now; since every thing
> Puts on the semblance here of sorrowing.
> Sick is the Land to'th'heart; and doth endure
> More dangerous faintings by her desp'rate cure.
> But if that golden Age wo'd come again,
> And *Charles* here Rule, as he before did Raign;
> If smooth and unperplext the Seasons were,
> As when the *Sweet Maria* lived here:
> I sho'd delight to have my Curles halfe drown'd
> In *Tyrian Dewes*, and Head with Roses crown'd.
> And once more yet (ere I am laid out dead)
> *Knock at a Starre with my exalted Head.*[13]

[13] The reference to his "Curles," etc., may be taken as a gloss on the

It will be evident that the court of Charles I, which had formerly excited attack from even the most dedicated Cavaliers for many of its vices, now becomes something to be longed for. (One is still arrested, however, by Herrick's percipience, if one may stress his words differently: "And Charles here *Rule,* as he before did *Raign.*") The moral issues that Jonson never ceased setting forth yielded to a national crisis for a whole class of the most literate people in what they had once been able to call the realm. Anything would be better than the Rule of the Saints, taking moral pretexts for acquiring the lands of the Royalists.

"The bad season" was that of the Cavalier winter. Why this image was chosen for hard times should be evident to anyone who has weathered a European winter without adequate heat. It was also an old commonplace, "Hyems tristis," as Manuzio put it in his *Flores.*[14] Whether the politicized Anacreontic grasshopper was an emblem of man's uncertain state or an object lesson, he and the hard weather became prime poetic concerns. In "Farwell Frost, or welcome the Spring," Herrick observes that outside a warmer season has come.

> Fled are the Frosts, and now the Fields appeare
> Re-cloth'd in fresh and verdant Diaper. . . .
> What gentle Winds perspire? As if here
> Never had been the *Northern Plunderer*
> To strip the Trees, and Fields to their distresse,
> Leaving them to a pittied nakedness.   (1-2, 9-12)

---

frontispiece of *Hesperides.* The golden age longed for is that of the return to rule by Saturn (Charles) and the return of Astraea (Justice); see Virgil, *Eclogues,* IV. 6.

[14] My edition is a Latin-French text, *Elegantiarum Aldi Manutii Flores* (Lille, 1675), p. 282. Manuzio had died in 1515, so that his commonplaces and elegancies served Europe from, as it were, the introduction of printing to the end of the seventeenth century.

We must pause here to consider the obvious and prepare for the uncertain. Obviously, and with very few exceptions, seventeenth-century poets do not simply describe "external nature"—a concept that was only dawning in English minds. Nature included man as well as God's lesser creatures; and between man and the other creatures were numerous resemblances and correspondences. This obvious fact should not excite us to say what resemblances and correspondences were being drawn (they being possibly legion), unless we have some particular form of evidence. And a great deal of poetry was in fact written between 1640 and 1660 in which a political or topical intent seems as certainly to be meant as to be certainly difficult or impossible to unravel.[15] In the passage just given, for example, "the *Northern Plunderer*" is Boreas. Is he also a figure, as some editors think, for the Scottish invasion early in the Civil Wars? Given the passage alone, surely it would be licentious, unscholarly, poor criticism to say so. But Herrick's last six lines must be considered.

> So when this War (which tempest-like doth spoil
> Our Salt, our Corn, our Honie, Wine and Oile)
> Falls to a temper, and doth mildly cast
> His inconsiderate Frenzie off (at last)
> The gentle Dove may, when these turmoils cease,
> Bring in her Bill, once more, *the Branch of Peace*.
>
> (17-22)

As before the political tenor of the seasonal imagery was subdued, so here the tenor all but excludes the seasonal imagery. Yet even that dove bears weightier freight than she appears to. It is the dove that flies back to Noah's ark, the ship of state (cf. Horace, *Odes*, i. xiv) after the tempest and flood.

---

[15] Lovelace's handsome but baffling poem, *The Falcon*, is my favorite example, but I have been unable thus far to understand the poem myself or to persuade my colleagues or students to undertake the task.

The flood is the disordered state, as *Cooper's Hill* (or Horace, *Odes*, I. ii. 1-2) shows us. Of these images, all of which are used repeatedly, the wind or tempest and the cold winter are those that most often provided Cavalier poets from Herrick to Cotton with an understanding of their plight during the Civil Wars and the Interregnum.

iv. Remedies of the Times

After the massive figure of Jonson had passed, the most distinctive feature of the Cavalier response to the times was retreat. Such is, after all, the usual action taken by those who are defeated but who need not surrender. Men like Davenant, Suckling, Cowley, and Waller (not to mention the frightened Hobbes) crossed the Channel. But most lay low and, on the evidence of their poetry, waited out the winter with wine and friends. We have just seen Herrick hoping for the return of a political spring, and we can easily recall Lovelace's poem, "The Grasse-hopper," in which (as in Herrick's poem), we first get description of the outer scene and then application and a movement within.

Few supporters of the King can have retreated so far inward as Henry Vaughan. In "An Epitaph upon the Lady Elizabeth, Second Daughter to his Late Majestie," he treats an event that occurred in September, 1650.

> Thou seem'st a *Rose-bud* born in *Snow*,
> A flowre of purpose sprung to bow
> To headless tempests, and the rage
> Of an Incensed, stormie Age.   (15-18)

The passage seems to lay before us so much that we have come to think characteristic of Cavalier poetry: the rose for the woman, the *discordia concors* of flower and snow (also suggesting the "red and white" of feminine beauty?), and the symptoms of the Cavalier winter. But the retreat turns to a triumph in what was termed apotheosis in the last chapter.

> These were the Comforts she had here,
> As by an unseen hand 'tis cleer,
> Which now she reads, and smiling wears
> A Crown with him, who wipes off tears.   (35-38)

Any reader will recall *Lycidas* and the union with Christ Milton envisioned, like Vaughan here (borrowed from Revelation, vii. 17). What keeps Vaughan's verses barely in the orbit of Cavalier poetry is his social sense of his audience, "here," where he and it stand together. But certainly he has moved farther away than Herrick or Lovelace from the times, and in *Silex Scintillans* "The Retreate" into the soul is very nearly complete, and the Cavalier mode is displaced by the Metaphysical. Some poems in the later published *Thalia Rediviva* do indeed look outward and enunciate their Royalism. "To the pious memorie of C. W. Esquire," for example, depends on the Cavalier images of "the dark," "storms and changes," "mists," "the insuperable stream of times," and "Ecclipses." In the Cavalier winter a friend has died. Every Cavalier poet of importance who survived into the years of the Interregnum seems to have a poem of this kind. Yet once again, we find that Vaughan moves to the very margin and almost loses his sense of this world by absorption in the next. Such mingling of modes frees every reader to classify or describe as he wishes. To me, the close of this poem is closer to the Metaphysical mode, unlike the poem just quoted, and in spite of its use of military imagery. To understand the differences (as I perceive them), let us compare an earlier passage on the life of Charles Walbeoffe, Vaughan's cousin.

> For though thy Course in times long progress fell
> On a sad age, when Warr and open'd Hell

Licens'd all Artes and Sects, and made it free
To thrive by fraud and blood and blasphemy . . .

(49-52)

That is second-rate Jonson, not second-rate Donne. On the
other hand, the ending catches an altogether different poetic
mode from such comment on the times.

Besides, Death now grows decrepit and hath
Spent the most part of its time and wrath.
That thick, black night which mankind fear'd, is torn
By *Troops* of Stars, and the bright day's *Forlorn.*
The next glad news (most glad unto the Just!)
Will be the Trumpet's summons from the dust.
Then Ile not grieve; nay more, I'le not allow
My Soul should think thee absent from me now.
Some bid their Dead *good night!* but I will say
Good morrow to dear Charles! for it is day.   (83-92)

That is true Vaughan all right, in one of his styles (we will
see in another chapter that it is not his only one), and it must
be termed Metaphysical. The social mode has yielded to the
private, as one radical symptom. In finding reason for a Cavalier retreat in Cavalier values, and in turning for succor to the
religious meditation of the Metaphysicals, Vaughan enacts
in this poem much of his own career as a poet, and he also
presents a microcosm of his age. The as yet but barely understood middle decades of the century brought the century's
peak of Hermeticism, other likely and unlikely "neo-Platonic" strains, Rosicrucianism, astrology, natural magic, and
inner light. The Cavalier need to retreat in those times may
be viewed in obvious economic and political terms. But it
may also be regarded as a version of the spirit of the age.
Depending upon the direction taken, the distance gone, and
the report after arrival, a given poet could start with Cavalier

assumptions and end with them, or like Vaughan, start with them and sometimes end with others.

Among such inward travelers with all manner of visa from the court of Apollo, there is preeminently Andrew Marvell. His critics have found it difficult to agree with each other on how to sort out the differing emphases in his enigmatic writing. Pierre Legouis put us all in his debt, first in French and then in English (*Andrew Marvell: Poet. Puritan. Patriot*) with his study of Marvell. No doubt attaches to his calling Andrew Marvell a poet. But what kind of patriot he was or was not, and what kind of Puritan, these are lingering questions. Most of us think some of his poems fully Metaphysical: "The Definition of Love," for example. Other poems seem Cavalier in motif and subject and theme but Metaphysical in operation: *To his Coy Mistress*, for example. Yet other poems forecast the public mode of the Restoration: *An Horatian Ode*, for example. But there is something distasteful to us all in cutting up a poet like a cheese, and "if we would speak true, / Much to the Man is due." So let us briefly consider *The Garden* and *Upon Appleton House* (I am not sure that brief consideration of either will seem to knowing readers to take us very far, but they will concede that brevity is novel in this area). *The Garden* derives from sources drawn upon by a number of Metaphysical poets and very much explores the Metaphysical private mode. And yet curiously enough, the poem was made possible by the same events that sent Lovelace to consider the ant and Herrick the frost. Fairfax's withdrawal, even retreat, to his country estate responded to the stern march of events as much as did Herrick's looking for friends when he was turned out of his Devonshire parish. And in writing *Upon Appleton House, to my Lord Fairfax*, Marvell concerned himself with very un-Metaphysical subjects, subjects that had been the concern of poets from Ben Jonson to Marvell himself. He writes about how Appleton House had got the way it was, that is, about how England had got the way it was. Having looked very closely indeed; having appreciated and even flattered Fairfax to a nicety; having talked of war, slaughter, and flood; Marvell on the whole wonders whether, the times being as disordered as they are, the Fairfax estate to which they have retreated might not be said to represent an order difficult to find outside it. In this last stanza but one of the poem, he is continuing his address to Appleton House itself.

"Tis not, what once it was, the *World*;
But a rude heap together hurl'd;
All negligently overthrown,
Gulfes, Deserts, Precipices, Stone.
Your lesser *World* contains the same.
But in more decent Order tame;
You Heaven's Center, Nature's lap.
And Paradice's only Map.   (761-68)

Here we discover something of Ben Jonson's address to Penshurst in "Those short but admirable Lines" distinctive of Marvell. Here is Donne's favorite figure of the corresponsive microcosm. Here is the kingdom-Eden figure as it runs from Shakespeare to Pope. Here is not "a rude heap together hurl'd," however, but the Cavalier strategy of turning inward to find that order and integrity that has disappeared without. Such is the strategy indeed, but the brilliant, untraceable tactics of such a retreat, as the very last stanza alone would show, make Marvell the Xenophon of Cavaliers. He has the Cavalier vision of those days and presents it in a shaded green to a great, troubled, Puritan lord and general. Within the year Marvell was tutoring Cromwell's daughter and, according to Milton, offering himself for service to the state. Marvell baffles most of us. Whether he followed some consistent line through most inconsistent times, or whether he too was buf-

fetted off the straight line by the storms of the day—these
matters invite very different interpretation. For present purposes, however, we may accept this much: Marvell understood the Cavalier assumptions and drew from them without
taking oath to them any more than he did to kingship. Indeed it is curiously easier to say what he was not. The senses
in which he was not a Puritan and in which he was not a
Cavalier emerge very well from John Aubrey's report. Aubrey
knew Marvell; and we know that wine and friendship are
Cavalier emblems.

> He was of middling stature, pretty strong sett, roundish
> faced, cherry cheek't, hazell eie, browne haire. He was in
> his conversation very modest, and of very few words: and
> though he loved wine he would never drinke hard in com-
> pany, and was wont to say that, he would not play the
> good-fellow in any man's company in whose hands he
> would not trust his life. He had not a generall acquaint-
> ance. . . .
> He kept bottles of wine at his lodgeing, and many times
> he would drinke liberally by himself to refresh his spirits
> and exalt his Muse.[16]

Everyone finds something of the amphibian, or griffin, in
Marvell. A bibulous Puritan or a solitary Cavalier—each
seems to deny nature and all laws of criticism enacted these
three centuries. Perhaps once we have explored the middle
decades of the century more wholly, we shall understand the
Marvells and the Cowleys better. At this juncture, we may
say with that safety conferred by what is obvious, that retreat during the Cavalier winter brought many shifts and
shadings of poetic style.

One of the freshest voices to sing Cavalier songs during

[16] *Aubrey's Brief Lives*, ed. Oliver Lawson Dick (London, 1949),
p. 196.

that winter was that of Charles Cotton, who appeals to us almost in excess of his abilities, because of his naturalness and sanity. Sane as he is, however (and truly he has some particle of that divine Chaucerian fire), his method of directness makes him a bit hard to understand in a century of indirectors. At least, he can show us what such other poets as Vaughan and Marvell cannot, a pure Cavalier subject: *Winter* a poem of 53 stanzas of double tetrameter couplets.

> Hark, hark, I hear the *North* Wind roar,
> See how he riots on the Shoar;
> And with expanded Wings out-stretch,
> Ruffels the Billows on the Beach. (1-4)

No other seventeenth-century poet writes of natural scenes with such consistently descriptive effect. Of course, like the rest, Cotton must move to a higher order than the unfigurative, and although the storm comes on, wrecking a ship, what follows is a kind of personification. "Winter and all his blust'ring train" come on invasion, not as guests (93-104). There is nothing for it but to retreat.

> Fly, fly; the Foe advances fast[;]
> Into our Fortress, let us hast
> Where all the Roarers of the *North*
> Can neither Storm, nor Starve us forth.
> (149-52)

The military imagery calls for attention, and so too the success in defying winter.

> Let him our little Castle ply,
> With all his loud Artillery,
> Whilst Sack and Claret Man the Fort,
> His Fury shall become our Sport. (205-208)

How far does the military imagery justify our suggesting a military or political meaning? Those who do not recall numerous such winter poems might think the coincidence of winter and war merely accidental. Those who have encountered such images as "the *North*," winter, storms, and battle again and again in unquestioned political contexts will likely think the political reading inevitable. After all, did it not survive to 1660 and that once enormously popular poem by Robert Wild, *Iter Boreale*? Does not Cotton soon refer to "the plotting Presbyter"? (In line 210.)

In my view, Cotton's winter campaign is political, but the poem as an entirety qualifies both less and more for political reading. It qualifies less by reason that the political references are confined (as in most such poems) to a brief section of the poem. In our total experience they play only a momentary part, and in the sequence of our reading their role is not great unless we choose to isolate them, which then makes much of the political reading depend upon our discovering just such a reading. But a political interpretation qualifies more by virtue of evidence subtle as well as specific. The specific comes in the brief political references and in various strands of imagery. The subtle evidence is that of the retreat from threat. Such withdrawal differs from that of the Metaphysicals, who never wanted (or pretended never to want) a place in the sorry world. The Cavaliers find in their retreat a measure of defense in a bad season. And their little world provides, not so much the ecstasies of love profane or divine, but the joys of friendship. To them, solitude would be all but rude compared to such sweet society.

Jonson had preached a Virtue Militant. After 1645 or so, the strategy became that of defense. But just as subtle reasons compel us to find a reaction to political events in poems of retreat, so too are there often subtle reasons for shifting the balance to say that a poem is really about friendship or nature

or activities other than those primarily political. Cotton's
verse address "To my dear and most worthy Friend, Mr. Isaac
Walton," for example, begins "in this cold and blust'ring
Clime" and goes on to expect a sunshine that would make
possible friendship and fishing. He and his friend would

> . . . think our selves in such an hour
> Happier than those, though not so high,
> Who, like Leviathans, devour
> Of meaner men the smaller Fry.　　(37-40)

Here we surely have a political or social glance in a poem
whose emphasis lies elsewhere. In other poems, like Lovelace's "Snayl" we can be certain of political meaning, and in
that lovely poem by the same poet, *The Falcon*, we are led
to suspect such additional dimensions without being able to
offer much more evidence than the man who disliked Dr.
Fell. In spite of their many protestations, I have strong
doubts that many of the people given to writing about such
things retired to the country simply because they preferred
it. Bacon lived the life of retirement *after* his trial, and pro-
tests from men like Cowley only seem to indicate that political or economic reasons were after all uppermost. (Restoration comments about lack of support of Cowley, Butler, *et al.*
only confirm this view.) But, and this needs emphasis in
treating of human beings, once "politics" or "social circum-
stances" had led a man to retreat to his (or to my Lord Fairfax's) estate, he could find other reasons for enjoying the
country.

One of the lasting benefits of the Cavalier winter seems to
me to be precisely the discovery, or rather the immortalizing,
of the joys of English country life and its *convivialities* in the
strictest sense. When, with the Restoration, the Cavaliers
bounced back to London to become Pope's "Mob of Gentlemen," they took years to adjust to the changes then occur-

ring. Social changes soon brought disappointment, and a new life required poetic talents other than those that had made survival possible during their winter. Very many of the old Cavaliers (Davenant, Denham, Waller) survived into the Restoration. New men of similar stripe emerged: Dorset, Sedley, Roscommon, Mulgrave, and such. Although the old and the new worked together to maintain courtly values, we observe some fundamental changes. The best Cavalier poet to emerge after the Restoration was John Wilmot, Earl of Rochester, who occupied a narrower human isthmus even than that Pope would describe for man. Rochester's allegiance to the Cavalier line is evident in his lyrics and his life. He took Suckling's negligent code seriously, with the result of a tragic self-waste and a nihilism from which Suckling himself with all his tact hardly escaped. Rochester also differs from Cavalier poets in his social origin in the nobility. The best Cavalier poets held a rank no higher than a baronetcy; and Jonson had of course been apprenticed to a bricklayer. Similarly, the best Restoration poets came from the gentry: Dryden, Etherege, Oldham, Wycherley, Shadwell, and Congreve. Yet a stronger sign of movement away from the "courtly makers" of the sixteenth century comes with the emergence of Milton, son of a money-lender, or a Bunyan, son of nobody at all. Other reasons for our discounting the possibility of a Cavalier revival after their winter exist, but perhaps the most important one is that the new stage for poetry will be found in the public forum. After the Restoration, some poets continue to write of retirement, but such writers tend to carry aspects of Metaphysical poetry rather than of Cavalier poetry into that late season. The Cavalier poetry of retirement, in brief, lost its essential character after the winter had passed and as poets came alive to their more active role in national life. For a Marvell it meant a coarsening of his poetry. For a Waller it meant a realization of possibilities that he and Denham had glimpsed earlier. For a Milton or a Dryden, it offered a chance to be themselves.

Since retreat in the hard times was a remedy for those who came as it were between Jonson and Rochester, we may now turn to other remedies available to Cavaliers at all times. Three such remedies—ceremony, myth, and civilization— perhaps make up but one basic remedy, although some convenience resides in the distinction. Of ceremony one may safely say that it would be difficult to find another Cavalier value less integral to twentieth-century concerns. Most people today can only wonder over the rankling, continued debates as to what language Adam and Eve spoke, what kind of music should be sung in church and by whom, or whether stained-glass windows suited primitive Christianity. No doubt all of us today would think a Lord Mayor's dinner or a two-hour Independent sermon ceremonious indeed. But ceremony in fact provides one of the most effective means of distinguishing poetic Cavaliers from other poets. There is no ritual in Donne's poem, "The Flea."

The masques show how much ceremony was prized. In non-dramatic poetry, the Cavaliers adapted, sometimes ironically, the ceremonies of the Petrarchanists (the blazon, the complaint, the "persuasion," and even the palinode); and they devised their own, especially along the lines of Roman models.[17] Such devising set out with the aim observed in their religious ceremony, of steering between the whorish finery of the papacy (or Elizabethan sonneteers) and the slatternly bareness of Calvinism. We recall that the lady in Herrick's "Delight in Disorder" is got up in fine dress. She wears a

[17] In *Poetry and the Fountain of Light* (London, 1962), H. R. Swardson devotes two chapters to the general subject of the relation of Christian and pagan elements: "Herrick and the Ceremony of Mirth" and "Marvell and a New Pastoralism." The fullest studies, apart from those of echoes of classical poets, have been Robert Deming's; see ch. ii, n. 7, above.

"Lawne" about her shoulders, has a dress with lacy runs, cuffs with ribbons, and a "Crimson Stomacher," while beneath it swish a petticoat of some body and shoes tied with strings. But no jewels or cosmetics are mentioned. The ceremony involves manipulating such lively dress, and it resides in the one-by-one parallel itemizing. Marvell showed (with his usual witty qualifications) his understanding of the rites for the Coy Mistress.

> An hundred years should go to praise
> Thine Eyes, and on thy Forehead Gaze.
> Two hundred to adore each Breast:
> But thirty thousand to the rest.
> An Age at least to every part,
> And the last Age should show your Heart.
> For Lady you deserve this State;
> Nor would I love at lower rate. (13-20)

The glories of our blood and state, as Shirley put it, may be shadows rather than substances, but they mattered. Marvell's wit does not destroy the positive statement, any more than does Suckling's humor in "Loves Siege" and A Ballade. Upon a Wedding, two fine poems. Perhaps today only a Roman Catholic would easily understand that ceremonies are so important that they do not usually need to be taken with solemnity. They make worship possible, but should not stifle it.

Herrick's poetry celebrates the Cavalier rites more persistently than does that of the other Cavaliers, except perhaps Habington, who remained, I fear, a recusant in spite of Jonson's reformation. Of Herrick's ceremonial gestures, that adapting Roman customs to his own purpose probably strikes attention first. The level of adaptation in the "Larr" poems hardly reaches that of the humanists' Christianizing of the classics. "Go where I will," he says, "thou luckie Larr stay

190

here, / Warme by a glit'ring chimnie all the yeare."[18] Without a sense of the value of ceremonies, or a knowledge of the extent to which pagan festivals survived into the seventeenth century, it would be difficult to understand how a Devonshire clergyman of poetic inclinations could write about the Roman *lares*. But when Herrick declared

> I sing of *May-poles, Hock-carts, Wassails, Wakes,*
> Of *Bride-grooms, Brides,* and of their *Bridall-cakes,*[19]

he chose just those elements that Puritans denounced as "carnal." It will be remembered that Bunyan had a severe struggle with himself until he could leave off enjoying church bells. And to oppose something to Herrick's argument, we may choose the carnal delights that Christian had to reject at Vanity Fair:

> Houses, Lands, Trades, Places, Honours, Preferments, Titles, Countreys, Kingdoms, Lusts, Pleasures, and Delights of all sorts, as Whores, Bauds, Wives, Husbands, Children, Masters, Servants, Lives, Blood, Bodies, Souls, Silver, Gold, Pearls, Precious Stones, and what not.[20]

The differences of opinion between men in the seventeenth century often seem to defy our sorting out, but on this score of ceremony the age did the work for us. Although not many poets might follow Herrick the whole distance and write of "The Court of *Mab,* and of the *Fairie-King,*"[21] he could claim wide assent for such various pagan, perhaps Celtic, elements as remain in *Corinna's going a Maying.*

[18] Herrick, "To Larr," ll. 9-10.
[19] Herrick, "The Argument of his Book," ll. 3-4.
[20] Bunyan, *The Pilgrim's Progress,* ed. James Blanton Wharey and revised by Roger Sharrock (Oxford, 1960), p. 88.
[21] Herrick, "The Argument of his Book," l. 12.

A deale of Youth, ere this, is come
Back, and with *White-thorn* laden home.
Some have dispatcht their Cakes and Creame.

(45-47)

In a sense Herrick often redeemed his Celtic paganism by lightly classicizing it, and his classicism by lightly Christianizing it. Similarly, he redeemed love-making by having it lead directly to marriage (49-50). One of Herrick's longer poems, *The Hock-cart, or Harvest home* (55 lines) seeks to create the quality of country life by presenting it in terms of a rural festival. The element of ceremony will be found in the repetition of certain words (e.g., "Come . . ." or "Some . . ."), and a certain "Devotion" appears in the blessing of crops and in drinking. Anyone with a decent edition will know that Herrick draws on Tibullus, *Odes*, II. i and can see by comparison how the classical, pagan spirit has been acclimatized. Herrick's debt is real, but his poem is very English. Dedicated "To the Right Honourable Mildmay, Earle of Westmorland," it addresses the "Sons of Summer, by whose toile, / We are the Lords of Wine and Oile" (1-2). The message simply says, Drink today, because tomorrow you must work:

And that this pleasure is like raine,
Not sent ye for to drowne your paine,
But for to make it spring againe.   (53-55)

I can see many virtues and interests in this. How interesting that the *carpe diem* urging should enter the social realm. How much we can prize an accurate and unvarnished explanation. But the rigid stratification of society and the unblinking affirmation of "paine" for the hinds and "Oile" for the gentleman is too much for me. (How much more generous

Jonson's *To Penshurst* is.) My distate does not extend to
Corinna; the rituals employed by Herrick in his vain effort
to win her are to me the timeless ceremonies and decencies
of the heart.

The best religious poetry of the Cavaliers seldom rates being called their best poetry, but it does emphasize the importance of ceremony. Herrick's "Ceremonies for Christ-
masse" is to be sure in the *Hesperides,* and it is no more
Christian than Jonson's *Christmas his Masque.* Not that I
think with Bunyan that it is sinful to enjoy oneself. "Ceremonies for Candlemasse Eve" (the feast celebrating the
Virgin's presentation of Christ in the Temple) concludes
with ceremonies suggesting very well the way in which observation of rites is a way of dealing with the times.

> Green Rushes then, and sweetest Bents,
> > With cooler Oken boughs;
> Come in for comely ornaments,
> > To re-adorn the house.
> Thus times do shift; each thing his turne do's hold;
> *New things succeed, as former things grow old.*
> > (17-22)

There is much to commend in ceremonies growing from the
natural cycle of the seasons or the cycles of marriage, birth,
and death, because recognition of such fundamental cycles
provides a ritual that affirms facts of life that our powers of
abstraction often conceal.

In *His Noble Numbers,* Herrick's religious verse is by turns
gnomically practical, paraphrastic of the Bible, and prayer-
like. Above all there is the ceremony of the Anglican Church
(to which are joined Roman and folk rites as well). In "The
Parasceve, or Preparation," there is even a preceremony for
that of Holy Communion. In like wise, Ben Jonson's "To
the Holy Trinitie" takes on the character of a liturgy, a prayer

193

for Trinity Sunday that might be used each year. This poem, and two hymns, appear at the beginning of *Under-Wood* with a larger title, "The Sinners Sacrifice." The sense of ceremonial usage is present in the very titles and emerges clearly in the third stanza of the poem to the Trinity.

> All-gracious God, the *Sinners Sacrifice*.
> A broken heart thou wert not wont despise,
> But 'bove the fat of rammes, or bulls, to prize
> > An offring meet.   (9-12)

Behind these verses lie passages in the Psalms.

> The sacrifices of God are a broken spirit: a
> broken and a contrite heart, O God, thou wilt not
> despise.   (LI. 17)

And in Psalm, 1. 13-15, Jehovah tells Israel,

> Will I eat the flesh of bulls, or drink the blood of
> goats?
> Offer unto God thanksgiving; and pay thy vows to
> the most High;
> And call upon me in the day of trouble: I will
> deliver thee, and thou shalt glorify me.

Of course "To the Holy Trinitie" also allows Jonson to incorporate almost a ritualistic treatment of the mystery of the Three that are One. No one would seriously argue that such poems rival in poetic quality the best Metaphysical religious verse. That said, one may also say that there is also a difference in the kind of experience created. Jonson very well knew what private prayer is and Herbert very well knew the details of divine service. But the center of Jonson's religious poetry does not lie in private meditation any more than does that of Herbert in public worship. Such distinctions emerge clearly in a letter written by that stout Cavalier, James Howell, while imprisoned in the Fleet (1 September 1645; II. lxvii). He had received some "Sparkles of Piety" in manuscript and was pleased to find in it "many most fervent and flexanimous Strains of Devotion." But he objected to a work consisting only of "supplicatory Prayers" and urged his correspondent "to intersperse among them some eucharistical Ejaculations, and Doxologies, some Oblations of Thankfulness." Above all, "we should not be always whining in a puling petitionary way (which is the Tone of the Time now in fashion)." Howell opposed what came to be called Presbyterian and Puritan canting, favoring praise and such ritualistic devotion as is implied by words like "eucharistical," "Doxologies," and "Oblations." With such differences of religious emphasis in mind, and with the Cavalier love of ceremony in secular matters well established, we can see why their religious poetry moved to the highest point it attained when dealing with ritualistic subjects, however flexanimously.

Cavalier poetry is ceremonial in very many respects. The lyric poetry is usually divided like hymns into discrete stanzas, and the occasions celebrated by poetry offered the opportunity to perform social rites.

> For Lady you deserve this State;
> Nor would I love at lower rate.

Now everyone knows that the next word in *To his Coy Mistress* is "But." And that time is the problem. Yes, the ritual will not defeat time, at least not in that poem by Marvell. Other poems may be more optimistic, but in any event the important thing is that the poetry is, so to speak, as unconsciously ceremonial as it is consciously absorbed with time. One of the finest of Cotton's intuitions reveals his sense of how much of human life can be understood in diurnal or seasonal ceremonies. Although *Summer Quatrains* and *Winter Quatrains* show as much, perhaps his finest perceptions

195

are found in the sequence of *Morning, Noon, Evening,* and *Night Quatrains.* The very passage of the day, time itself, becomes a rite defeating time, as the *Night Quatrains* show.

XV.

Now in false Floors and Roofs above,
The lustful Cats make ill-tun'd Love,
The Ban-dog on the Dunghil lies,
And watchful Nurse sings Lullabies.

XVI.

Philomel chants it whilst she bleeds,
The *Bittern* booms it in the Reeds,
And *Reynard* entring the back Yard,
The Capitolian Cry is heard. . . .

XIX.

The Sober now and Chast are blest
With sweet, and with refreshing rest,
And to sound sleeps they've best pretence,
Have greatest share of Innocence.

XX.

We should so live then that we may
Fearless put off our Clotts and Clay,
And travel through Death's shades to Light;
For every Day must have its Night.

No one would think Cotton the greatest poet of the century, but at moments like those in the first two stanzas quoted, he seems the most original, for truly in him that "catholic decorum" that belongs to the century as a whole becomes a capacity not only for bringing together what seems discrete but also for bringing things together for their own intrinsic worth, without seeking some grand correspondence between them. And the last two stanzas, which conclude the diurnal quatrains, take us to that other world of Charles Cotton, the world he shared with Izaak Walton, and yet without violating in the series of quatrains the sense that he possesses a total vision. Life, like the day itself, becomes a ceremony, and death or night but take us to a lasting light beyond. The rites of the day well lived come to take on far greater import than we had suspected at first. They, and of course the unspoken Anglican Christianity in this and so much Cavalier poetry, offer man hope of a salvation within the very teeth of time.

If ceremony proved a means of discovering an order in time and the times, it will not be difficult to show that mythmaking closely followed ceremony. There is a sense in which *Corinna's going a Maying* and *The Hock-cart* present English nymphs and satyrs in bucolic tales. Jonson's "Hymne" in *Cynthia's Revels* (IV. iii), arguably the most perfect lyric of a century of perfect lyrics, addresses that *"Queene* and *Huntresse,* chaste, and faire"* in almost the only poem in English that combines something truly like classical quantity with usual English metrical stress. It does so in a way making Diana inhabit the world of classical myth, while at the same time allowing for her to represent a myth for Elizabeth. No casual hand has shaped this poem, and the congruity of its conclusion with that of Cotton's *Night Quatrains* should tell us that here we find something essential to seventeenth-century poetry and our own lives.

> Lay thy bow of pearle apart,
> And thy cristall-shining quiver;
> Give unto the flying hart
> Space to breathe, how short soever:
>   Thou, that mak'st a day of night,
>   Goddesse, excellently bright.   (13-18)

Jonson and Cotton, like Vaughan and Milton and Dryden, and indeed Christians of many persuasions knew why moving from darkness to light mattered.

The people that walked in darkness have seen a great
light: they that dwell in the land of the shadow of death,
upon them hath the light shined. (Isaiah, ix. 2)

The Christianizing of this prophecy (and Bunyan may be
included with the others named) led to differing kinds of human and literary experience, especially since Isaiah, ix. 1-8 is
the first lesson for matins on Christmas Day in the Anglican
service, and because the verse quoted provided the basis for
the Introit of the second Nativity Mass of the Roman Catholic Church. To the Cavaliers, it was less essential to bear
down upon the Christian element. The passing of day to
night to the new light provides a myth that need not exclude
Diana, Elizabeth, and biblical echoes. Cotton brings his elements onstage together; Jonson allows them to suggest each
other.

At times, in his masques and in *To Penshurst*, Jonson's
mythopoeic powers seize on motifs that we would naturally
associate with myths of the golden age or of the earthly paradise, often with a vision of plenitude representing God's providential care of His human creatures. The poem on the Sidney estate includes as well such details as dryads, muses,
sylvanes, and satryrs, with local legends to boot (10-18). But
the total ordered life borrows from all these potentially mythic elements (religion, classics, social hospitality, the earthly
paradise) to create what may be called Jonson's own myth.
To Penshurst is Jonson's version of the gardens of Alcinuous,
and if we gain the strong impression that he, like Odysseus,
preferred to go back to his own city, that is true, too, and in
other poems the city or the court could be the subject of other
visions. Carew's "To Saxham" clearly follows in Jonson's
topographical tracks and presents motifs that have grown
very familiar. Winter blows outside, while the fire burns within. Carew sounds many of Jonson's notes, but his synthesis

of such elements moves in the direction of that philosophiz-
ing symbolism that he draws upon so heavily for his masque,
Coelum Britannicum. I confess that "myth" really should
be restricted to certain forms and patterns of narrative, but
when food becomes a witty concordance of the four elements
(earth, water, air, and fire) and when that takes us to a ritual
of fire and our search for light in the dark, then something
more than mere imagery is involved.

> Water, Earth, Ayre, did all conspire,
> To pay their tributes to thy fire,
> Whose cherishing flames themselves divide
> Through every roome, where they deride
> The night, and cold abroad; whilst they
> Like suns within, keepe endlesse day.
> Those chearfull beames send forth their light,
> To all that wander in the night . . .   (29-36)

The inclusive term for such poetic effects may be left to the
reader recalling poems from Donne to Dryden, but especially
those we have now been considering.

When Carew wrote in his elegy on Donne that soon the
poetic "Libertines" would "Repeale the goodly exil'd traine /
Of gods and goddesses, which in thy just raigne / Were
banish'd nobler poems" (63-65), the prediction was safe:
they had never left. In an earlier chapter we observed how
much of the success of Carew's poem, *A Rapture*, depended
precisely on everything happening in "Elizium." And however much critical indigestion "myth" has caused, it provides
the wholesome sustenance of masques, as all would agree.
What we may think especially important about the use of
myth there reflects back on what I somewhat abusively
termed myth in topographical poems. That is, just as we have
observed the way in which ceremony presented a means of
admitting and dealing with time, so also can we see that

myth might deal with disorder by admitting it to a higher order. We now understand that Jonson devised his version of the antimasque from earlier variants of antimasque or ante-masque, and we can see very easily that he employs the counterform in numerous ways. But what he realized by mind or instinct was that the mythic order he sought to portray could best be earned, and indeed tolerated, by partial admission of a disorder to threaten it. This is no place to enter on discussion of the drama of the time, but one really must emphasize that the most devoted of Cavaliers had doubts about the actual court and about the possibilities of too sterile an order. Such doubts were sound in political terms and healthy for art. Many tragedies and tragicomedies of the time deal centrally with the vices of princes and the threat so posed to social order. Some playwrights (e.g., Webster, Middleton) sometimes went so far as to make the possibility of any real order merely a bond of hope between playwright and audience. But most "serious" plays admitted more or less disorder to the end more or less of proving the enduring order of providence in this world. Of course it takes no effort to show how the comedies of the time move, on their lower social level, from disorder to order.

To return to Cavalier poetry, however, it will be recalled that the dedicatory poem to the *Hesperides* is addressed "To the Most Illustrious, and Most Hopefull Prince, Charles, Prince of Wales." Here we find Tudor-Stuart myth unabashed: religious and astronomical imagery, correspondences including the sun emblem, and a veritable apotheosis. And yet Herrick could also write in "Bad Princes pill their People":

> So many *Kings*, and *Primates* too there are,
> Who claim the Fat, and Fleshie for their share,
> And leave their Subjects but the starved ware.

(4-6)

It would be difficult to summarize the objections to Charles I and Archbishop Laud more succinctly than that. And yet even here there must be at least an antimyth. The first three lines develop a simile: "Like those infernall Deities. . . ." It very much seems to me that Cavalier poets sought to admit on *their* terms aspects of disorder so that they might in conscience affirm the myth of a large, beneficent social order from which they in particular benefitted or hoped to benefit. When the terms on which disorder was to be understood were set by the regicides, the problem of the Cavaliers grew acute. But the tendency to what I have loosely termed mythmaking remained one of the most imaginative capacities, and assurances of order, enjoyed by these poets.

What above all the Cavalier poets possess as a means of protection from the disordered times was the conviction that their own individual integrity was part of a civilization comprehending God's true English church, the "laws and liberties" of England, and the rules and enjoyments of the classics. Fundamental political and even economic changes were underway in England of the seventeenth century. And we know that even when change brings real benefits it may seem disorderly and dangerous to those too immersed in affairs to perceive the directions being pursued. The Cavaliers were unshaken in their conviction that the worlds of poetry and art were on their side. Dryden's sentiment, "And never Rebell was to Arts a friend," was not wholly just, nor, some have thought, was Milton wholly fair to his famous artist, Satan. Milton allowed no one to possess greater devotion than himself to the Royalist cause: at the Court of Heaven. And good Cavalier that he was, there, he could express doubts that the Son was constitutionally of one substance with the Father, though he paid both all possible worship.

Most of the elements making up "civilization" in Cavalier poetry have already appeared in the preceding discussion. I shall close this chapter with a few remarks on some prevailing attitudes toward Cavalier poetry and some examples from Herrick. One view, one more common than might be admitted, holds that Cavalier poetry possesses the Elizabethan gift for song but that it admitted more artifice. No one who has come so far in this book is likely to be deceived into thinking that I believe that, any more than I believe that Jonson's prime distinction as a poet is some hypothetical classicism of polish and restraint. But one purpose of this chapter certainly has been to stress how repeated motifs, attention to ceremonies and rites, and something like myth-making contribute to a view of civilized order making life worthwhile. Another strange view, still held in gross distortions, turns Waller and Denham into the villains who usher in "the Restoration and the eighteenth century."[22] It is perfectly obvious that the public mode triumphant after the Restoration shared more with the social mode of the Cavaliers than with the private of the Metaphysicals. Without trying to do battle with real or imaginary Dick Minims among our critics, however, I shall bring forth Robert Herrick as a poet who sometimes sounds notes of a proto-Augustan kind. Among the seven poems titled "Upon himself" is one beginning with two lines that employ what may be called the Augustan metaphor:

> Come, leave this loathed Country-life, and then
> Grow up to be a Roman *Citizen*.[23]

[22] Among recent studies offering valuable correctives to older views, see Paul J. Korshin, "The Evolution of Neoclassical Poetics," *Eighteenth-Century Studies*, II (1968), 102-37, who is especially interesting on Cleveland and Waller.

[23] Herrick, "Upon himself," ll. 1-2. Maynard Mack discusses aspects of such appropriation of the Augustan values to England in " 'Wit and Poetry and Pope': Some Observations on His Imagery," in *Pope and his Contemporaries: Essays presented to George Sherburn*, ed. James L. Clifford and Louis A. Landa (Oxford, 1949), pp. 20-40, especially pp. 36-39.

One can say that much of Pope and Dr. Johnson is instinct
in those two lines, although one cannot credit them a place
among Herrick's finest poems. Somewhat more must be said
on behalf of a better "Augustan" poem, "His returne to
London."

> O fruitfull Genius! that bestowest here
> An everlasting plenty, yeere by yeere.
> O *Place*! O *People*! Manners! fram'd to please
> All *Nations, Customes, Kindreds, Languages*!
> I am a free-born Roman; suffer then,
> That I amongst you live a Citizen.
> London my home is.   (7-13)

The very syntax comes to seem Latin. Indeed, Herrick is
standing on tiptoe rather than taking flight, agreeable though
that posture appears. And we observe the irony of the model
world of this country poet turning out to be London rather
than Devonshire, whereas a major model for Ben Jonson of
Westminster was not London but a country estate.

Herrick, who has figured so largely and so well in this chapter, has yet another "Augustan" poem that requires no apologies whatsoever, except to him for its not being better known.
Now, alas he does not return to London but shed "His tears
to Thamasis" as he departs from him.

> No more shall I reiterate thy Strand,
> Whereon so many Stately Structures stand:
> Nor in the summers sweeter evenings go,
> To bath in thee (as thousand others doe.) . . .
> Never againe shall I with Finnie-Ore
> Put from, or draw unto the faithfull shore:
> And Landing here, or safely Landing there,
> Make way to my *Beloved Westminster*:
> Or to the *Golden-cheap-side*, where the earth
> Of *Julia Herrick* gave to me my Birth.   (3-6, 11-16)

The wit and beauty and force of such lines move me deeply, and not of course for their being proto-Augustan, or for telling us of the "civilization" of the Cavaliers, but for their being poetry and Herrick. And in words like "state" in the following passage, or its imagery of flood, or its consciousness of proper ceremonies—in such things we have a fresh version of many of the tropes (and themes) of order and disorder treated throughout this chapter.

> Keep up your state ye streams; and as ye spring,
> Never make sick your Banks by surfeiting.
> Grow young with Tydes, and though I see ye never,
> Receive this vow, so *fare-ye-well for ever*.   (23-26)

The ordering principle of poetry—whether imagination, life transformed, imagery, feeling, or pressure of mind—operates here, accounting in its way for possible disorders as well, including those two last, parting and death.

LOVE

There is an honest love . . . of a most
attractive, occult, adamantine property,
and powerful virtue, and no man living
can avoid it.

—Burton

IT COULD BE argued with some force that no century of English literature produced finer, or at least more various, poetry than did the seventeenth on the subject of Burton's "honest love," as well as on certain other varieties. No critic living can avoid the subject, and nothing, surely, would seem more necessary or indeed simpler in a book on Cavalier poetry than to write a chapter on love. There are Amoret and Aurora, Celia, Castara, Sacharissa, Corinna, Lucasta, Althea, and Ellinda; there are also Dorinda, Chloe, Flavia, Celinda, Laura, and (with my apologies to any ladies I have forgot), Zelinda.[1] In spite of the abundance of what Herrick called his "fragrant Mistresses," nothing so central to Cavalier poetry as love has turned out to be so difficult for me. That is, one may indeed speak of individual coy mistresses with some confidence. One can observe how important are concepts of the good life and of time to poems on love (as I have already done). And one can specify certain conventional topics and motifs (as I shall do shortly). But when one comes to those topics that have made the greatest noise—"Platonick love," for example—one is embarrassed by the lack of evidence. Or again, it has proved difficult for people to agree

---

[1] Propertius, II. xxxiv. 85-94 authorizes such catalogues.

on the extent to which real experience enters Cavalier love poetry. The Corinnas are frequently dismissed for lacking the reality of, say, Donne's women in the *Songs and Sonnets*. One kind of "reality" is of course the felt poetic pressure, and that at least can be discussed on literary grounds. But some commentators on Cavalier love poetry appear also to feel that it lacked sufficient contact with realities of seventeenth-century experience.

i. The "Powerful Virtue" of Love

Such an allegation requires historical, sociological, and psychological evidence to sustain or refute it. Such evidence is precisely what we have lacked. If it is a satire on recent historical scholarship of the first half of the seventeenth century that its brilliant practitioners have only brought obscurity, it is a scandal that we literary scholars have done so little.[2] To pose a question that may seem utterly mad, how old is Herrick's Corinna? Now, just such a question has exercised the learned editors of Donne over his elegy (if it is an elegy), "The Autumnall." Donne's editors agree that either possibly or certainly the poem is addressed to Lady Magdalen Herbert. There is dispute over the year in which Donne was born, and considerable disagreement about when the poem was written. Perhaps in 1600, perhaps in 1609.[3] If we accept the

[2] A notable exception is L. C. Knights' *Drama and Society in the Age of Jonson* (London, 1937; Peregrine Books, 1962), which still excites readers, in spite of its rather dated political assumptions of the R. H. Tawney sort. Everyone hopes for a breakthrough with the help of the present Cambridge group of sociologists.

[3] In *The Poems of John Donne*, 2 vols. (Oxford, 1938), II, 62-63, Sir Herbert J. C. Grierson argued that "The Autumnall" was written *ca.* 1608-1609; in John Donne, *The Elegies and The Songs and Sonnets* (Oxford, 1965), pp. 146-47 and Appendix C, Dame Helen Gardner argued that the poem was written about 1600. See also *The Complete Poetry of John Donne*, ed. John T. Shawcross (New York, 1967), p. 401, for a brief note.

recipient and the date of 1600, the poem was written to a lady of thirty-five or thirty-six by a man of twenty-seven (if Donne was in fact born in 1573); if it was written in 1609, she was forty-four or forty-five and he about thirty-six. And shortly before the supposed date in 1609, she married a person not half her age. Something in nature prompts me to imagine that a poet might respond differently to a woman nearly ten years his senior than to one his own age or younger.

The problem we face comes down to this: what ages did the Cavalier poets fancy their male wooers and their ladies to be? That is, what kind of relation between men and women do they develop in their poems? These are hard questions. The parish records are said to suggest twenty-eight as the mean age for a woman to marry. On the other hand, Waller (b. 1606) started to woo his Sacharissa when she (Dorothy Sidney, b. 1617) was seventeen or eighteen and he about twenty-nine. John Evelyn (b. 1620) married in 1647, he being twenty-seven and his wife Mary (b. ca. 1635) eleven or twelve, and they deeply attached to each other. Pepys's wife was a few years older. Whatever the sociological evidence, however, the literary convention appears to entail that the men poetically wooed their Sacharissas when they were about twenty-five and their ladies about sixteen or seventeen. It will be recalled that even at a later period of the century a girl was thought ready for love at fifteen.[4] If, at the same season that he has urged a girl in the bloom of her beauty to yield to him, our Cavalier writes verses to a male friend on the superiority of friendship to love, may it not be possible to explain such things humanly? May it not well be that he had more in common with a man of his own age and education than a wom-

[4] See Dryden's "Song for a Girl" in *Love Triumphant*, v. i. (1694). She is fourteen; and the shepherd Karolin who sings "Though I am young, and cannot tell" in Jonson's play, *The Sad Shepherd*, i, is probably about the same age; Jonson's song provided the model for Dryden.

an perhaps barely more than half his age and perhaps scarcely able to read?[5] Such a hypothesis does not conflict with the tone of Cavalier love poetry, but let us say that where Corinna's age is concerned hypothesis alone is possible. Perhaps prose will afford us firmer ground than the ages of fictional heroines.

Bacon writes "Of Love" as on all subjects, like a lord chancellor, and like most judicial pronouncements, his are true as far as they go. "The stage," he says, but he might have said the poet, "is more beholding to Love, than the life of man." And he distinguishes the kinds of love and their results. "Nuptial love maketh mankind; friendly love perfecteth it; but wanton love corrupteth and embaseth it."[6] Whether that perfecting "friendly love" be love as considered here or friendship as considered in the next chapter may well be doubted. And sometimes we may wonder which kind of love is being treated in poetry. Habington's love for his Castara, Lucy Herbert, was almost "Platonick" in style but "nuptial" in aim and outcome. In his lovely elegy, "An Exequy," Henry King addresses his dead wife as "Friend." Of course the poets were more aware than we that in Latin *amor* and *amicitia* are so like, and that, as Cicero was translated in 1691, "certainly *Love* (from whence the Name of *Friendship* is deriv'd in Latin) is the first and strongest tye of our Affections."[7] Male sexual imaginings play a considerable part in Cavalier as in other poetry, often making it difficult to distinguish between "friendly love" and "wanton love." Carew's *Rapture* comes at once to mind as a paean to wanton love somehow

---

[5] From Donne's letters, it appears that Anne More, his wife and a niece of Lord Keeper Egerton, was illiterate. Similarly, the orthography of Lady Elizabeth Dryden must have been learned in great haste.

[6] Bacon, *The Works*, ed. James Spedding, *et al.*, 7 vols. (London, 1870-1872), VI, 397-98.

[7] Anon., *Cicero's Laelius. A Discourse of Friendship. Together with a Pastoral Dialogue Concerning Friendship and Love* (London, 1691), p. 21. *De Amicitia*, VIII. 26.

not really so wanton, off there in "Elizium." Similar problems rise with other poems and poets. What range of male or female experience is implied, and how really amorous is the urging of Corinna to go a-Maying or of the Coy Mistress not to be coy? In fine poetry, the sexual passion (like other passions) takes of its own accord a considerable freight of other meanings. What it could not seize by itself was often given it by a Herrick or a Marvell.

But we have forsaken our lord chancellor, whose comments of course proceed by precedent, even when not so acknowledged. "I know not how," Bacon reflects with a commonplace, "but martial men are given to love: I think it is but as they are given to wine; for perils commonly ask to be paid in pleasures." The commonplace here very well suits what was formerly given such names as the Cavalier spirit. Such a conception of drinking, fighting, wenching Royalists induced enraged Parliamentarians to stigmatize their opponents as Cavaliers. Good King Charles's Golden Days acquired their gilt, as we have seen, by the passage of time, and by the end of the century Dryden could write of the early part that it was "A very Merry, Dancing, Drinking, / Laughing, Quaffing, and unthinking Time."[8] Colonel Richard Lovelace was certainly not "unthinking," but he was one of Bacon's "martial men," one "given to love . . . as . . . to wine." Even in those poems most familiar from the anthologies, however, a Lovelace could combine so finely and humanely a political with an amorous idealism that his poetry seems to remain forever fresh. For example, the first three stanzas of "To Althea, From Prison" provide us with a paradigm of that old-fashioned entity, the Cavalier spirit: love, wine, and Royalism. The point we shall see is one that I have been able to extract out of the lord chancellor on behalf of my client: despite

---

[8] Dryden, *The Secular Masque*, ll. 39-40; however, Momus, not Dryden, says this.

some circumstantial evidence, Cavalier love poetry is not truly guilty of "wanton love."

I.

When Love with unconfined wings
　　Hovers within my Gates;
And my divine *Althea* brings
　　To whisper at the Grates:
When I lye tangled in her haire,
　　And fetterd to her eye;
The *Gods* that wanton in the Aire,
　　Know no such Liberty.

II.

When flowing Cups run swifty round
　　With no allaying *Thames*,
Our carelesse heads with Roses bound,
　　Our hearts with Loyall Flames;
When thirsty griefe in Wine we steepe,
　　When Healths and draughts go free,
Fishes that tipple in the Deepe,
　　Know no such Libertie.

III.

When (like committed Linnets) I
　　With shriller throat shall sing
The sweetnes, Mercy, Majesty,
　　And glories of my King;
When I shall voyce aloud, how Good
　　He is, how Great should be;
　　　Inlarged Winds that curle the Flood,
　　　Know no such Liberty.　　　　(1-24)

When a man can sing his caged linnet notes so sweetly while expecting death, we have to admit the truth of the song. Or if it is not true, then it should be. And once again we have found ourselves arrived at an idealism that carries an unaccustomed variety of poetic transcendence.

Such transcendence is most remarkable for springing out of uncertainty. It would be difficult to exaggerate the surprise one feels today on first encountering the nervousness felt by writers of the seventeenth century on the subject of love. Whether Burton's *Anatomy of Melancholy* provides the readiest explanations depends on whether one can find them, but surely no one wrote on the subject of love melancholy at greater length.[9] We must begin with those kinds of love that the age felt safe with.

> Besides this Love that comes from Profit, pleasant, honest, (for one good turn asks another in equity) that which proceeds from the law of nature, or from discipline and Philosophy, there is yet another Love compounded of all these three, which is *Charity*, and includes piety, dilection, benevolence, friendship, even all those virtuous habits; for Love is the circle equant of all other affections . . . and is commanded by God. (III. i. 3)

In one of his rare quotations of an English poet, Burton illustrates what he terms "the greatest tie, the surest Indenture, strongest band" between human beings—friendship—from Spenser.[10] We must understand (in spite of my chapter divisions) that in much of the seventeenth century love and friendship are, although distinguishable, constantly thought of together.[11] The distinguishable, concupiscible

---

[9] Burton, *The Anatomy of Melancholy*, ed. A. R. Shilleto, 3 vols. (London, 1920), III, 1 ff. Later citations are to parts and sub-parts.

[10] Spenser, *The Faerie Queene*, IV. ix. 1-2 (correcting Burton's citation); Burton, III. i. 3.

[11] See the anonymous volume cited in n. 7, above. From his research into friendship in certain literary traditions, David Latt has drawn my attention to the fact that in the Elizabethan miscellanies "friend" is a common term for one's beloved, and of course friends often speak of

love did not enjoy a very good press in the seventeenth century, and Burton finds no dearth of evidence (mostly from earlier moralists) to castigate it. But, and this is the important thing, he, like the poets from their quite different view, was able to transcend his suspicions. He identifies other kinds of love, specifying

> that *Heroical Love*, which is proper to men and women, [and] is a frequent cause of melancholy, and deserves much rather to be called burning lust, than by such an honorable title. There is an honest love, I confess, which is natural, . . . a strong allurement, of a most attractive, occult, adamantine property, and powerful virture, and no man living can avoid it. (III. ii. 1. 2)

Burton has come as he usually does, the long way round, to that which seems to have been missed by those theorists given to short cuts: "and no man living can avoid it." Poets, true poets, as Jonson would say, make that aspect of the human condition theirs. In a love elegy, Jonson himself characteristically has the poet arrogate sway over this realm of "occult" and inevitable love.

> Let me be what I am, as *Virgil* cold,
>   As *Horace* fat; or as *Anacreon* old;
> No Poets verses yet did ever move,
>   Whose Readers did not thinke he was in love.[12]

So much for the poet's right to be himself and his responsibility; but the wonderful tone is beautifully sustained over several lines.

> Put on my Ivy Garland, let me see
> Who frownes, who jealous is, who taxeth me.

---

loving each other. Otway's *Venice Preserved* welters in such interassociations.

[12] Jonson, *An Elegie*, ll. 1-4 (*Under-Wood*, XLIV).

Fathers, and Husbands, I doe claime a right
   In all that is call'd lovely: take my sight
Sooner then my affection from the faire.
   No face, no hand, proportion, line, or Ayre
Of beautie; but the Muse hath interest in:
   There is not worne that lace, purle, knot or pin,
But is the Poëts matter: And he must
   When he is furious love, although not lust. (9-18)

I will not enter large claims for all the rest of this elegy, but
in these lines will be found some of the most perceptive
literary criticism of the seventeenth century. A little thought
will show how far in advance of his century Jonson is in
raising questions about the poet's concern with the universal
in terms of detail, in grasping the similar but different versions of an experience like love in poetry and in life, and
in understanding (or boasting) of the creative powers possessed by the poet. Love must indeed have a "powerful virtue" to increase Jonson's understanding so far. But his fascinating discussion merely uses love as a point of departure.
To understand the full range of what love meant to the
century, from "burning lust" (in Galenic lore, the heart was
source of heat) to cool "Platonick" ecstasy, we shall have to
take Burton's longer route through more prose.

Two short stories of ancient origin and Continental popularity reveal, especially with their fascinating commentaries,
the "Power of Love & Wit": *The Ephesian and Cimmerian
Matrons*.[13] To take the stories first—in *The Ephesian Matron*,
the matron mourning her dead husband in a charnel house
takes up with a passing soldier; *The Cimmerian Matron* concerns a yet more sordid tale of a wife, a husband, a bawd, and

---

[13] The two stories, with augmentations in excess of their length, were
published together in 1668. There is no need to enter on the European
popularity of the stories, sources, and bibliographical details, but I shall
presume the usual association with Dr. Walter Charleton.

a soldier. The stories are of lasting interest only by virtue of
the commentaries for which they very strangely provide the
pretext. At its 22nd page, *The Ephesian Matron* is inter-
rupted for 46 pages of varying commentary and argument
about love, whereafter the tale continues for another dozen
pages or so with minimal intrusion. In a prefatory epistle before *The Cimmerian Matron* the person signing himself
"P. M. Gent." addresses his "dearest friend," the "Author
of the *Ephesian Matron*." P. M. defends his friend against
charges of writing "*a studied* Satyr *against* Women."[14] He
defends him for the good reason that defense is needed. His
own tale is repulsive: and yet, and this is what makes us shake
our heads about the century, P. M. adds a lengthy section of
commentary of an *idealistic* kind.[15] What these writers have
to say about Platonic love will be a useful starting point.

Dr. Walter Charleton (whom I shall presume to be the
author of *The Ephesian Matron*) concludes with an essay,
"Of Platonick Love." To the best of my knowledge, this is
the first original such essay in English, and like the mainstream of seventeenth-century poetry it rejects the "Platon-
icks." Charleton knows a prime source of the doctrine of
Platonic love, that is of neo-Platonic interpretations of it, the
Symposium, and he also knows the criticisms made by Lucian
and Petronius. But he insists on the "vast disparity . . . betwixt the Platonique Love of the Ancients [which in any case
he has doubts about], and that of Modern Puritan Lovers."[16]
How the Puritans got in there may seem altogether obscure,
but it is of course their reputation for hypocrisy and sensual-
ity that is involved. Certainly, in Charleton's view, "this
Platonique Passion is but an honourable pretence to conceal
a sensual Appetite."[17] In this again he sides with the poets of

[14] P. M., sig. G2ʳ.
[15] P. M., *The Cimmerian Matron*, pp. 32-77.
[16] *The Ephesian Matron*, p. 67.
[17] *Ibid.*, p. 68.

the century. If what Plato said about the aim of love in the *Symposium*—"to impregnate [Alcibiades] with that knowledge and those Virtues, with which [Socrates'] own Mind was pregnant"—then, Platonic love "is the same with our *Charity*."[18] Here in little we discover the usual points made against Platonic love. It is intellectual ("love of the mind" is the usual phrase), incorporeal; such love cannot exist between man and woman; when disinterested love is found, it turns out to be nothing other than Christian charity.

Other streams flowed underground, however. What to name these shadowed waters is a question not yet answered to general satisfaction. Neo-Platonism, mysticism, perennial philosophy, spiritualism, kabbalism, these are some of the terms in use for some time. More recently, "syncretism" has emerged as the general descriptive term, and although the word does not seem particularly shapely, the meaning comes closer to the fact. Seventeenth-century writers, on the other hand, commonly used such standard rhetorical terms as "allegory."

> Nor are they meere Inventions, for we
> In th' same peece find scatter'd *Philosophie*
> And hidden, disperst truths that folded lye
> In the dark shades of deep *Allegorie*,
> So neatly weav'd, like *Arras*, they descrie
> *Fables* with *Truth*, *Fancy* with *Historie*.[19]

Since there will always be people who find shadows luminous, we can only assume that "*Allegorie*" retained its attraction from, say, Ficino to Blake with consistency in its appeal and little consistency or fixed form to its ideas. A few facts illustrate our difficulty in sorting out the ideas that might under-

[18] *Ibid.*, pp. 63, 65.
[19] Henry Vaughan, "Monsieur Gombauld" (*in Olor Iscanus*), ll. 41-46.

lie the supposed cult of Platonic love at the court of Queen Henrietta Maria. Effectively, Plato was not published in England until the eighteenth century. On the other hand, in the mid-1580's there occurred a sudden flurry of interest in "*Allegorie*," for which the readiest symptom is the publication of seven separate titles by Giordano Bruno in two or three years. For that early a period, such frequency astonishes. Then, in an often-quoted remark from 3 June 1634, James Howell spoke of there being at court "a Love called *Platonic Love*, . . . a Love abstracted from all corporeal gross impressions, and sensual Appetite, but consists in Contemplations and Ideas of the Mind, not in any carnal Fruition."[20] It is this phenomenon that has received a great deal of mention and very little explication. We shall return to it after pointing to some other symptoms. Certainly, the 1650's saw a revival of "*Allegorie*." *The Divine Pymander of* "Hermes Trismegistus" was first published in England in 1650, and again in 1657. As long before as 1487, Pico della Mirandola had published his *Commento sopra una canzona de amore da H. Benivieni*. But only as late as the 1651 edition of Thomas Stanley's poems did this appear in English under the title of "A Platonick Discourse Upon Love." Stanley's choice cannot be faulted. In this, his single important work written in the vernacular, Pico gives an admirably brief, clear account of the chief neo-Platonic doctrines and symbols.[21]

Some of the same virtues of comprehensiveness in brief

[20] Howell, *Epistolae Ho-Elianae*, 11th edition (London, 1754), p. 255 (I. vi. 15). If any early seventeenth-century poet of stature deserves to be called a poet of Platonic love, it is no doubt Lord Herbert of Cherbury. But his most Platonic poem is not on love, and one of his poems labeled "Platonick Love" is not Platonic. Further, his resemblances to Spenser and Donne, in these poems, hardly qualify him for consideration here.

[21] Pico's commentary is a marvel of lucidity and succinct comprehensiveness. In Crump's edition of Stanley, the "Discourse" comprises but 401 lines of type. Stanley reprinted the essay in *The History of Philosophy*, 4 vols. (London, 1655-1662), II, 94-118, after his description of

(a virtue never more to be prized than in neo-Platonic writings) will be found in P. M.'s little essay, "The Mysteries and Miracles of Love" at the end of *The Cimmerian Matron*. I cannot say why this and Charleton's writings have been so ignored. With the third part of *The Anatomy of Melancholy* and Stanley's translation of Pico they provided English readers with serious treatments of love. P. M. knows Ficino's views. He looks with favor on "Mysteries of a Divine Fury."[22] He recounts the story of *Cymon and Iphigenia* to illustrate the doctrine of love as education—the "Scoole of love," in a contemporary phrase.[23] He is concerned with *"Extasie"* and *"Transmigration."* He considers each *"Individual* in love . . . [to be] thenceforth a *Number"* or a Hermaphrodite. And so he goes through the topics that scholars have read into earlier English poetry on the usually undemonstrated assumption that their English poets were reading books published abroad, usually very much earlier. With P. M. we stand on firmer ground, and he speaks at last of

> *Platonique* Love, or generous *Charity*; the delight whereof consisting likewise in the exercise of ones *power* or ability to enrich the understanding of another, and impraegnate his Mind with the seeds of Virtue . . . *Platonique* [love] depending solely upon the Mind, whose powers are perpetual, is therefore calme, of one equal tenour, and everlasting.[24]

Plato's philosophy. Because Stanley presents traditional accounts, often in simple translation, his discussions of classical philosophies possess special value to students of the century.

[22] P. M.'s phrase is somewhat reminiscent of Giordano Bruno's *De gl' Heroici Furori* (London, 1585). P. M.'s observations were published 83 years later, in 1668, the year after *Paradise Lost*. I shall have occasion at the end of this chapter to reflect on how little the "sensibility" of either the Restoration or the earlier part of the century has been understood.

[23] Dryden translated *Cymon and Iphigenia* with the same moral, and a larger view, in *Fables*, 1700; the phrase, "Scoole of Love" is Cotton's.

[24] See "The Mysteries and Miracles of Love" in *The Cimmerian Matron*, pp. 73-74.

Most seventeenth-century poets simply did not write as if they believed this. On the other hand, most would not have gone so far in the other direction as Dr. Charleton, who refused to allow a distinction "betwixt *Love* and *Lust*."

> Those unprejudicate Enquirers, who have searched deep enough into the Origine and essence of that desire of Conjunction in persons of different Sexes, or the Appetite of Male and Female to each other, which is generally understood to be *Love*, (for, we are not now upon consideration of *Amity*, or *Friendship*) will not be easily perswaded, that there is any so great dissimilitude or Disparity betwixt them.

Some such reductive suspicion continued to lurk in the English mind, making it suspicious of the high-flown "Platonic" doctrines, and certainly led English poets to write far more "anti-Platonicks" than paeans. Charleton specifically rejects "the *Platonique* sect,"[25] in which P. M. must be numbered. And so at last, we come to the question: where is this love poetry "depending solely upon the mind" to be found?

My answer will displease some people. In my view, the concern with "Platonism" far exaggerates its importance to Renaissance poets, and certainly it exaggerates their understanding of Plato. Indeed, there was no understanding of Plato in any full sense between the Academy and the great German classicists of the last century. As for neo-Platonism, so called, there are numerous inky or bright streams, and we need another A. O. Lovejoy to insist upon our thinking of a plurality of neo-Platonisms. One such is classical or pagan, deriving from later stages of the Academy, whether in Greece or Rome. Another, again pagan, becomes Christian by commentaries like Macrobius's on Cicero's *Dream of Scipio*.

[25] Charleton speaks of "*Love* and *Lust*" in *The Ephesian Matron*, p. 48, and he rejects "the *Platonique* sect" on pp. 49 and 62-69.

Early Christian neo-Platonism of probably the most important kind derives partly from Cicero through St. Augustine,
before Augustine had learned Greek; and then from the im-
position of Augustine's theology upon Plato after he learned
Greek in his mature years. Later kinds of neo-Platonism derive heavily from Ficino's translation of Plato, which was apparently the only complete Latin translation to circulate and
the one that is repeatedly reprinted. The British Museum
Catalogue shows that Ben Jonson and Sir Kenelm Digby
owned Continental editions of Plato and so, we may believe,
did some other learned poets. As we might expect, that
nursery of Cambridge Platonism, Emmanuel College, has
one of the richest holdings of neo-Platonic books and editions of Plato. Most of its approximately 15,000 books with
imprints before 1700 were, however, acquired during the
seventeenth century, a third very late in the century from the
non-juring Archbishop Sancroft. As has been said, "syncretistic" lore of many kinds was revived in mid-seventeenth
century when Hermetica became attractive to some (and
when that prolific Cambridge Platonist, Henry More, chas-
tised the Hermetic Thomas Vaughan for "preposterous and
fortuitous imaginations"). By the 1670's and 1680's the work
of mid-century bore numerous fruits in fields of religion,
philosophy, literature, and science. Some symptoms are the
relations drawn between one kind of neo-Platonism and Cartesianism (a natural enough connection) and the first appearance in English printing of Plato's works in any language
in 1673, 1675, and 1683. But the works of Plato are simply
not readily available in England until the eighteenth century,
and understanding of Plato through the Greek and without
the veil of Plotinus and Ficino comes only with the Nine-
teeth century. Such are the indisputable facts.

Of course what is true in gross many be false in individual
cases , and facts require interpretation. To put it very briefly,

I think that only the most strenuous interpretation can lead to the belief that Henrietta Maria instructed the lords and ladies of her court in matters philosophical. The simplest explanation is usually the best, and surely it is that the good queen imported *preciosité*, that it was called "Platonick love," that she chose to sponsor such a "love of the mind" at court, and that the poets begged leave to write their "Anti-Platoniques."

Let us return briefly to James Howell, whose evidence has seemed so important. About three years after reporting on the outbreak of Platonic love at the court, he reports (3 February 1637):

> F. C. Soars higher and higher every Day in Pursuance of his *Platonic* Love; but *T. Man* is out with his, you know whom; he is fallen into that Averseness to her, that he swears he had rather see a *Basilisk* than her.[26]

Obviously "*Platonic* Love" should be translated "court armours." *Preciosité*? Almost certainly. Platonic love? If one wishes to call a mere affair so. No wonder Charleton termed so-called Platonic lovers "modern Puritan lovers," concupiscent and hypocritical.

To dismiss the importance, to poetry, of "Platonic love" at the Caroline court is not to dismiss varieties of neo-Platonism from Cavalier poetry nor to say that the subject is without importance. In view of information given above, it is not surprising, for example, that when he celebrated Venetia Digby after death, Jonson should use one of the commonest neo-Platonic concepts in seventeenth-century England. In his "Picture" of her body, he presents it as the macrocosm of the world in a very interesting hexaemeral or creation passage; and of course, the next poem is on the mind. As for Henry More's Psyche in his *Psychozoia*, so for Lady Digby: the

[26] Howell, *Epistolae Ho-Elianae*, p. 326 (II. xvi).

world (her body) is the garment of her mind. It is also true that if we pick our way carefully, we can find paths of neo-Platonism that take us some distance toward clarity and even toward Plato as the century wears on. In his very illuminating *History of Philosophy* (4 vols., 1655-1662), Thomas Stanley gives a much clearer and in some ways more accurate picture of Plato's philosophy than had existed in England. It is symptomatic of a new spirit of philosophical rigor (in some quarters, that is), that Stanley begins with a return to the chief available Roman source, Cicero, in whose *Academica* (i. v) there is a three-part summary of Plato's beliefs. Stanley translates the passage and, in what follows, he at least attempts discrimination, even to the unheard-of point of separating off "Alcinous" (or Albinus Platonicus) into a separate section (pp. 56-93), giving the credit or blame to the single source. Moreover, he adds to that in yet another separate section (pp. 94-118) Pico's "Platonick Discourse" (that had appeared in the 1651 edition of his *Poems*, as has been observed).

Some of the evidence for the popularity of syncretistic thought in mid-century has been given. The fact that such traditional sources of "*Allegorie*" as "Hermes Trismegistus" were then finding their way into print in England at this time surely counts for far more by way of proof than the sort of gracious supposition usually offered us. Rather than pursue such writers here, however, I shall merely mention the emergence of the Cambridge Platonists and Cartesianism, as well as the fact that for some years those two streams of thought seemed quite compatible, as students of Henry More have shown. The evidence for the increasing understanding of Plato in the Restoration will be found in the frequency with which he is mentioned, especially by poets like Milton and Dryden, and partly in downright explications. John Norris, the one consistent Platonic poet of the century outside Henry More, published his *Collection of Miscellanies* in London in 1687. Internal evidence suggests that the poems included were written during the few years leading up to publication. Most of the volume consists of prose, however, and among the essays is "Another Letter . . . , concerning the true Notion of *Plato's Ideas,* and of *Platonic Love.*"[27] Norris rejects as false four earlier conceptions of what Platonic love is, arguing instead that it has as its object "Contemplation and Love of God, whom he [Plato] calls the *Idea of Beauty.*" But because of the "too sublime and refined excellency" of this *"Idea of Beauty,"* Plato "recommends to us . . . *a Method of Ascent,* which is from loving the Beauty we see in Bodies, to pass on to the Love of the Beauty of the Soul, from the Beauty of the Soul to the Beauty of Vertue, and lastly from the Beauty of Vertue . . . to the immense *Ocean of Beauty.*"[28] Many other things in Norris would show how much more adequate an understanding of Plato was now available. (Secular verse, usually of low intensity, poetically and platonically, was also written by The Matchless Orinda, Mrs. Katherine Philips, and her circle.) And Norris also demonstrates how a kind of Christian Platonic poetry was possible. In "The Exchange" (p. 128), unhappy Corydon swears that he will never love again. But he follows the speaker's advice to "Take bright *Urania* to thy Amorous breast"; and if that may seem too passionate, he adds, "And let thy *sensual* love commence *Divine*" (ll. 9, 12). I believe it true to say that "The Exchange" is the sole poem by Norris that touches on the existence of the female sex. "The complaint of Adam turn'd out of Paradise" (pp. 112-13) does not even mention Eve. No wonder he would dedicate a poem "To Dr. More" (pp. 89-91), and no wonder, too, that the great poets of the century felt obliged to be less pure.

[27] Norris, *A Collection of Miscellanies,* pp. 435-45; I have reversed roman and italic usage.
[28] *Ibid.,* p. 443.

Much remains to be studied, and especially in mid-seventeenth-century poetry. We may suppose that the learned poets of the century—Donne, Jonson, Milton, Marvell, and Dryden—had acquaintance with certain varieties of so-called Platonism. Readers on the lookout for "the dark shades of deep *Allegorie*," as Vaughan put it, will find them in some of his poetry—and in some of John Ogilby's *Fables of Aesop Paraphras'd in Verse*. At the end of *Comus* Milton introduces celestial Eros, and John Norris also talks about that exalted idea in his "Letter." Traherne and Marvell moved through such shades, and one can also find in them William Hammond. Full-scale views of man will be found in the tedious neo- or Cambridge Platonism of Henry More's *Song of the Soul*—and also in the typologies of Dryden's *Hind and the Panther*, and in the *musica mundana, musica humana,* and perhaps numerological symbolism of "A Song for St. Cecilia's Day."[29] But "neo-Platonism" appears to be as much a reality in Cavalier love poetry as the philosopher's stone was in the alchemy of that day. No doubt desirable, but exceedingly difficult to discover.

John Norris is a genuinely neo-Platonic poet, as is Henry More. Whether either is a genuine poet, however, may be doubted, and so may their inclination to the Cavalier mode. I have dwelt on them and on the so-called Platonic element, because the one "historical" thing about Cavalier love poetry has been supposed to be its cult of Platonic love. Perhaps by laying that ghost I shall save others the time and frustration I have spent. My skepticism about the whole topic will have been evident, and it is founded on what I think are unshak-

---

29 In the portion of *The History of Philosophy* devoted to Plato, Stanley treats such things as the sphere and circle emblems for perfect order (II, 70-71) and numerological symbolism for world harmony (II, 72-73). Other such "emblematic" or "syncretistic" lore can be found in his translation of Pico's essay (II, 94-118) and of course in sections on other philosophers.

able assumptions that philosophy is a serious business distinct from the practice of literature in many radical respects, that literary students are arrogant in seeking to treat quickly (and often with small Latin and less Greek) what centuries of philosophers and divines labored over with such difficulty, and that even in a century of learned poets, we cannot expect poets to possess philosophy in the way that philosophers do. Moreover, I do not think that "neo-Platonism" is a subject suspending normal rules of evidence, and if in one of its protean guises it is supposed to be a "source," then the transmission must be shown. Finally, and this brings me full circle, too many people think a presence of "neo-Platonism" is a guarantee of poetry (let them read Henry More), and that if a poem shows two lovers united, we automatically have concern with the androgyne, the hermaphrodite, the one and the many, numerological circularity, *concordia discors,* and such wonderful things. I shall not argue that all neo-Platonic writers are fuzzy-minded, long-winded, and all but irredeemably prosaic. But I shall ask whether "neo-Platonism" is really more poetic than, say, those touches or large measures of Calvinism shared by Donne, Jonson, Herbert, and Milton? Is it really more poetic than the Bible? Or, to say it all, is "neo-Platonism" really more satisfying on the score of what human love is like than Cavalier poetry?

The one thing that seems worth salvaging from the old question of historical "fact" about a cult of "Platonic love" is what may be adapted from the phrase, "love of the mind." Evidence does exist, in all manner of writings and in the poems themselves, for an increase in psychological awareness during the seventeenth century.[30] Burton is the readiest ex-

---

[30] See H. M. Richmond, *The School of Love* (Princeton, 1964), pp. 212 and 249-50 especially. Richmond is perhaps the first critic to have stressed sufficiently this major shift, and the importance of other such shifts, in the seventeenth century.

ample, devoting a whole portion of his work to love melancholy; but the "case of conscience" was as familiar to divines
of the time as are neurosis and psychosis to our physicians today. The love poems by Waller and Jonson and Cotton given
in the first chapter show this interest in psychological process,
especially in the perceiving lover, and they would also show
by comparison how great an advance was generally made over
the kind of understanding possible to most sixteenth-century
poets. (Shakespeare the sonneteer is an obvious but rare exception.) This rich vein was mined by seventeenth-century
love poets of all schools of love, but the psychological sterling
appears most obviously in Suckling. He could declare that
"There's no such thing as that we beauty call," and although
such a motif can be traced to classical writers, the next line
sounds a new note: "it is meer cousenage all."[31] Again, Suckling can say in the same poem, " 'Tis not the meat, but 'tis
the appetite / makes eating a delight" (17-18). Whereas
Donne took and refashioned the conceited aspects of Petrarchanism, Jonson and the Cavaliers refashioned and revital-
ized the psychological conventions of Petrarchanism. Everyone knows that Donne shows a mind in motion, and everyone can be quickly taught that conceits will be found in the
Cavaliers:

> and if I like one dish
> More then another, that a Pheasant is;
> What in our watches, that in us is found;
> So to the height and nick
> We up be wound,
> No matter by what hand or trick.[32]

---

[31] Suckling, "Sonnet II," ll. 9-10. See Richmond, *The School of
Love*, ch. v, "An Historical Perspective," pp. 224-77, for excellently detailed treatment of this and related motifs.

[32] Suckling, "Sonnet II," ll. 19-24. Lest someone jump to stigmatize

But Suckling will not be mistaken for Donne, nor Donne for that inventor of cribbage. By the time we get to Rochester, we observe that the stress upon the operation of the male (and sometimes the female) psychology has stretched personality to the breaking point.

The whole century has often seemed more modern than it is because of the division of mind in many men where matters religious or political were concerned. It was a rare major poet who did not change his mind on one or the other, or both, of these matters at least once. And the poetic direction encouraged by "syncretistic" thought parallels that Horatian movement inward, that centripetal transcendence that we have seen in one guise or another in Cavalier poetry. It is therefore no surprise that the love poetry should also turn reflective, self-observing, psychological. Some characteristic features of that process in the love poetry will be dealt with shortly, but in this, the most speculative or hypothetical section of the book, a premise may be suggested for further study.

The inward movement, which marks seventeenth-century poetry in kinds other than the Cavalier, and then in ways other than the Cavalier, surely received special emphasis during the Cavalier winter. After the defeat and then the execution of Charles I, nearly every poet of significance either turned in on himself (Marvell included) or more or less ceased to write poetry (as perhaps did Milton). The gain in new possibilities for poetry was much to be prized, but it became increasingly difficult to hold poems together, at least when the poems concerned love. The forms loosen, either in the sense of prosodic experimentation with what Cowley

---

Suckling as proto-Hobbesian mechanist for his conceit, let it be said that it is no more original with him than was Donne's compass conceit with him.

called the "Pindarique Ode," or what Cotton termed "Stanzes Irreguliers."[33] In their lyric poems, Davenant and Vaughan run on with great unevenness, and perhaps Traherne represents one extreme of dissolving form in his prose poetry, or Thomas Heyrick another in composing his bizarre semi-epic *Submarine Voyage* entirely in pindarics. The formal crisis following the triumph of the Metaphysical private mode and the Cavalier social mode came (it would seem) because both tended to move inward just far enough to risk balance. The new poetry swung back outward to the public mode. But it did not do so at once, at least not successfully. What distinguishes the new age is the reemergence of narrative, but the requisite skill was not quickly acquired. Cowley could not finish the *Davideis*, nor Davenant *Gondibert*, and some of their readers have had similar problems. Waller achieved success with a brief semi-narrative, "Upon a War with Spain, and a Fight at Sea," but he could not sustain a story in *The battell of the Sumner Islands*. Once again, we remark the extraordinarily complex and as yet insufficiently understood nature of mid-seventeenth-century poetry and thought. An age when Hobbes could intermit his excitement over geometry to write a Latin poem, *De Mirabilibus Pecci*, on the "wonders" of the barren Peak district of England, and when Charles Cotton could write *The Wonders of the Peak*, reversing the order in which Hobbes's wonders were introduced—such an age seems one in which anything might happen and in which any problem might occur. To define the range of love poetry before and during such a period may seem to require recollection of Marvell's formulation of a love "begotten by despair Upon Impossibility."

[33] Of course a Cowley or a Cotton would not be the fine poets they are without forms of control of other kinds, whether of the larger units such as Cowley's splendid *Essays*, or in the certainty of tone possessed by a Cotton.

ii. Some Conventions and Motifs

No convenient term exists for categorizing a range of conventions, motifs, formulas, and tropes deployed by Cavalier
poets in exploring their experience of love, real or imaginary.
After the queries and indirections of the preceding pages,
however, we may seek to follow Jonson and make simplicity
a grace and so arbitrarily distinguish a few simple elements
that may be termed schemes and motifs.[34] Among the
schemes one finds four familiar in criticism of sixteenth-
century poetry: the blazon, or itemizing of a lady's beauty;
the persuasion to enjoy, or seduction poem; the dialogue;
and the address to the coy or haughty mistress. I single out
these (and to them will add others) because of their being
so well known. And if I have mentioned their Tudor rather
than their Greek or Roman ancestry or their Italian and
French kinship, that is because we have heard enough of
foreign parts for the moment. What distinguishes the Cavalier blazon from its Tudor prototype is what is obvious
enough from the other schemes: its psychological emphasis.
Herrick's "Delight in Disorder" says as much in its title, or
in its first two lines:

> A sweet disorder in the dresse
> Kindles in cloathes a wantonnesse.

As the rest of the poem makes clear, the wantonness is partly
a result of the character of the woman, whose habits rather
than her dress are portrayed by the adjectives in phrases like
"an erring Lace" or a "cuffe neglectfull"; and partly a result
of the effect of such "wilde civility" on the male observer.

---

[34] Let me refer yet again to the detailed discussion of certain recurring
forms and ideas in ch. v. of H. M. Richmond's *School of Love*. (See
notes 30 and 31, above.) As I remarked in an earlier chapter, Richmond
also provides us in other sections of his book with the bases for new
valuation of poets like Waller.

The same tendency to represent thought and character by dress (which will not be found in Marlowe's description of Hero's whistling buskins or in his call to "Come live with me and be my love") will be found in great detail (ninety lines) in Cotton's "La Illustrissima." The superlative in the title reflects, even if in too generalized a way, a judgment by the male observer ("Oft have I lov'd, but ne'er aright," the poem begins), just as Cotton's detail lacks the tight spring driving earlier blazons.

Cotton affords another old motif that we recall from Campion and Jonson (and their Roman predecessor) in his "To Coelia": "When Coelia, must my old day set. . . ." That sun was setting on Catullus and Lesbia, but Cotton provides what is surely the mildest-mannered seduction poem of the century, even when it speaks of impatience.

> Yet think not (Sweet) I'm weary grown,
> That I pretend such haste,
> Since none to surfeit e'er was known,
> Before he had a taste;
> My Infant Love could humbly waite,
> When young it scarce knew how
> To plead; but, grown to Man's estate,
> He is impatient now.   (17-24)

The tone is delightful. The humor in the first four lines quoted depends upon the imperfection of an analogy, but we observe that it is the Sucklingesque love-as-appetite metaphor ("Hunger macht den besten Kuch"), and with the same examination of one's own psychology. We also observe, most markedly toward the end, how the not too complicated stanza form tends to loosen to no discernible purpose. It almost seems that Cotton has difficulty in making up his mind whether he is writing a song or whether he is writing a narrative of a lover's lifetime.

Dialogue in Cavalier love poetry usually consists of exchanges between characters with pastoral names. The tone of such exchanges varies from the erotic in Carew and lesser poets, to the frustrated in Marvell and a few others, and sometimes to the downright naughty in songs after the Cavalier manner by Rochester and others. What the pastoral element has to do with dialogue partly derives from tradition and partly provides assurance: jocund their Muses were, but their lives were chaste; or vice versa.

Another scheme, or perhaps two schemes, should also be mentioned, the dream of the beloved and what, after Carew, may be called the rapture. Since the dream of seeing the beloved, or perhaps the vision of her at night, may often culminate in a prospective or realized rapture, the two tend to merge. Much of this poetry is erotic, as the experienced reader of Herrick will recall. But such earlier writers as Herrick adjust their tone to convention so carefully that tradition combines with understatement to useful effect. After about 1660 some writers depart from convention to a more outspoken interest in the act of sex, sometimes to the end of tittilation, sometimes for psychological examination, and sometimes to express regret. But it will be clear to any reader that Jonson's moral rigor and fine compliment gradually alter into psychological exploration and occasionally into eroticism. The discovery of what seems to have been regarded as the naked truth of the passions, and the rediscovery of unvarnished Roman moods, sometimes recalls Donne's more than candid explorations in his elegies, and sometimes suggests the versification of male desires. The moralists share with the libertines, however, an interest in anatomizing the "affections" that prose writers as different as Robert Burton, Thomas Hobbes, Dr. Walter Charleton, and P. M., Gent. had taken as their province. Such psychoscopy of course recalls Montaigne, and it is altogether suitable that in 1685 Charles

Cotton should have translated the *Essays* anew, with one of his inimitable prefatory epistles.

Another scheme, if a seeming negative can be termed positively, is what one may term, with but slight exaggeration, the image-free love poem. Suckling, who is not my favorite as a human being among the major Cavalier poets, is nonetheless in some ways the most important for love poetry. That is, he seems to apprehend new tendencies, to write poems giving definition to such things in the air, and to establish them as canonical. Reading his love poems, it is not at all difficult to believe that he invented a card game for two players. We may take his three "sonnets": "Do'st see how unregarded now / that piece of beauty passes?"; "Of thee (kind boy) I ask no red and white"; and "Oh! for some honest Lovers ghost." These poems echo in the minds of poets to the end of the century, when Millament still has the first on her lips: "Natural, easy Suckling," indeed. These and other poems do stay in the mind, although perhaps less for lines arresting by themselves than for the casual attitude with which Suckling disassembles his own psyche, using as tools a minimum of images, and those not new. Perhaps Suckling's control, his mastery of language, form, and convention prove the crucial thing. (Cotton, whom I much prefer as a man, nonetheless may dissipate his effects by trying to be The Complete Gamester.) Suckling's image-free style also served to pass judgment, to make lyricism juridicial.

> There never yet was honest man
> That ever drove the trade of love;
> It is impossible, nor can
> Integrity our ends promove:
> For Kings and Lovers are alike in this
> That their chief art in reigne dissembling is.[35]

[35] Suckling, "Loving and Beloved," ll. 1-6.

Or again:

> There never yet was woman made,
>> nor shall, but to be curst;
> And oh! that I (fond I) should first,
>> of any Lover
> This truth at my own charge to other fools discover![36]

The two passages cannot be said wholly to lack imagery; and they do possess grace, force, and some complexity. The subdued contrast of the honest gentleman with love's tradesman touches real psychological and social depths in the century, as the comedy of the century shows, early and late. And in the second passage, the woman who discovers something like original sin for herself and who, she thinks, makes it embarrassingly known to others, is no small creation. Such "image-free" poems make up one kind of Cavalier love poetry strikingly at variance from Metaphysical styles. This kind proves enduring; at least it will be found in some of Prior's lyrics and in Goldsmith's "When lovely woman stoops to folly." Certainly, Suckling's music is often replayed during the rest of the century. In an "Ode," Cotton provides a late example.

> Was ever man of Nature's framing
>> So given o'er to roving,
> Who have been twenty years a taming,
> By ways that are not worth the naming,
>> And now must die of loving?　　(1-5)

The cadence of thought borrows from Suckling, and so too that psychological self-examination so characteristic of him in this style.

Another scheme that distinguishes certain Cavalier styles of love poetry from Metaphysical styles is that which I can best term semi-pastoral. I have occasionally alluded to this

[36] Suckling, untitled poem, ll. 1-5.

before, but the subject deserves a few connected sentences. One can at most call the scheme semi-pastoral because almost all the pastoral paraphernalia are dropped. Gone are the sheephooks, fleecing, sheep themselves, reed pipes, country cates and messes, and almost all the scenery; gone is the sense of perpetual spring; gone is the immunity from time. Gone indeed are those very elements that may seem to some tastes to clutter, or to others to adorn, the Elizabethan pastoral. The pastoral scene either all but vanishes from love poetry or takes on a new character, as in Marvell's Mower poems or Cotton's love poems with country settings. Even those well tried, if not always poetically true, pastoral devices such as song-contests are gone. One is reminded of a sign, which Matthew Arnold would have made much of, on the window of a vacant London shop: "Dryads have moved to Northgate, Leicester." The reduction is certain, and there may well be those who think that after so much has been subtracted, we have not semi-pastoral but hemi-demi-semi-pastoral. Or it may seem that what remains is pastoral in nature but no longer matters.

So it may seem. But we have evidence that the full-scale pastoral remained important in the funeral elegy. Falkland's poem opening *Jonsonus Virbius* is "An eglogue on the Death of Ben. Johnson, betweene *Melybaeus* and *Hylas*." There is pastoral Milton in English and pastoral Milton in Latin. There is Dryden "On the Death of *Amyntas*: a Pastoral Elegy." And by the end of the century a royal death brought out many, many small reeds to pipe their doleful notes and mournful ditties. The retention of pastoral detail in funeral elegies served a poetic purpose as buffer between the poet and the all but intractable fact he had to deal with. And I believe the removal of most detail from Cavalier love poetry also served a purpose, a purpose contrary to its retention in other poems: it enabled the poet to get to the experience far more directly, and to look far more easily into the characters and their motivation. Such capacities bear more directly on life and serve to deny the pastoral's traditional immunity to time.

The uses of pastoral depend in part on what of the pastoral is retained. Remarkably little was required in any one instance. A classical deity, unless perhaps the Sylvans or Pan, would not quite qualify. But a classical deity would almost be enough if there were also a flowery, warm outdoor scene. The use of a traditional pastoral device such as dialogue would help, and certainly we are home and dry with the mention of nymphs and swains. But above all, the presumption of semi-pastoralism emerges most readily when the characters are given pastoral names. Some such names, like Sidney's Stella, are newly made for their meaningfulness: Castara or Sacharissa, for example. Most come from very old pastoral inventories, or from analogy to such classical names. The reliance on names to create the semi-classical was a stroke of no doubt unplanned fortune, because the same names also suggested the Roman love elegy. Corinna might be a pastoral nymph, but everyone knew that the most famous of poetic ladies to bear that name had been Ovid's mistress (*puella, domina*), according to the *Amores*. Poems to Delia or references (by more poets than Milton) to Naeara's hair recall the Tibullan poems.

The pastoral is of course a highly sophisticated form. But by infusing it with elements of the love elegy, Cavalier poets managed to attain yet another kind of sophistication and also, by increasing their range and complexity, a new elasticity. This will be clear, I think from one of Waller's lesser poems, "A la Malade," published about the same time as *Lycidas*. Amoret has fallen ill.

> Ah lovely *Amoret*, the care
> Of all that know what's good or fair,

Is Heaven become our Rivall too? . . .
And as pale sicknesse does invade
Your frailer part, the breaches made
In that fair lodging, still more clear
Make the bright ghest, your soul appear.
So Nymphs ore pathless mountains born,
Their light robes by the brambles torn
From their faire limbs, exposing new
And unknown beauties to the view
Of following gods, increase their flame,
And hast to catch the flying Game.

(1-3, 21-30)

Or again, in another of Waller's poems from *The Workes* of
1645, "Chloris and Hilas" debate. She asks him to sing, offering birds as emblems of fidelity. His rejection of that virtue
concludes the poem.

Hil. Chloris, this change the birds doe approve,
Which the warm season hither does bring;
Time from your self does further remove
You, then the Winter from the gay Spring:
She that like lightning shin'd while her face lasted,
The Oak now resembles which lightning hath blasted.

(19-24)

In both poems, the pastoral element is slight but distinct,
and the general tenor hovers as much over an elegiac heroine
as a pastoral, providing a mixture set to purpose like other
characteristics of Cavalier love poetry, to examine human
thought and feeling. The same things are true of later poems.
In "The Surprize," Cotton recounts what happens "On a
clear River's flow'ry side" when he surprises his "glorious
Nymph" who has become amorous from reading the *Arcadia.*
And in "The Visit," the man, or swain, enters Castanna's

235

chamber and spends the night watching her asleep, unwilling to touch her. Very different, almost contrary, psychological states are sketched by the same semi-pastoral means.

In all these poems the pastoral, by being pastoral, admits a degree of advantageous control. The convention posits certain aesthetic and moral ideals for the poet to incorporate or counterbalance with other concerns. It can be argued that the pastoral idealized passion, providing the Cavaliers with control—or euphemism. It can also be argued, on the other hand, that the semi-pastoral offered a secure base from which exploration of amatory experience was possible. But I think it best argued that both possibilities were realized, that is, that sometimes the pastoral operated upon passion and that sometimes passion operated on the pastoral. Certainly much of the experience in Cavalier semi-pastoral love poetry is closer, say, to Tibullus's imagining of what life would be like in the country with Delia than it is like Drayton's *Idea* or Spenser's *Amoretti*. In the ambiguity of the Cavalier semi-pastoral, then, there was both control and escape. The control was inherent in the fixed aesthetic and moral values implied by pastoral—and such values are never very voluminous to the poets of any age. The escape was inherent in the psychological exploration of passion, almost always of male passion. Exceptions are almost always those of feminine regret or of dialogue poems, where both viewpoints are offered. Whether we have semi-pastoral or semi-elegy as a general category may indeed be open to question, and I have chosen the former term simply because it is clearer. In practice, a phrase or even a word will often prove crucial in setting the balance on the one or the other. When Jonson writes, "Come my *Celia*, let us prove, / While we may, the sports of love," the word "sports" signals the elegiac strain. When Herrick adds a few sweet natural details, we hear the pastoral strain.

Come, my *Corinna*, come; and comming, marke
How each field turns a street; each street a Parke
Made green, and trimm'd. (29-31)

At times, indeed, the balance may be so perfect and so pre-
carious that critics agree only with themselves. Marvell's
poem, *The Nympth complaining for the death of her Faun*
is just such a work. In it the Nympth complains not only of
the slaying of her faun but also of the falseness of Sylvio. The
moral grounding, erotic shading, and sophisticated design are
typical of what I have been calling the semi-pastoral.

The motifs of Cavalier love poetry deserve more detailed
examination than I can spare. (My pages of notes on the tradition of roses in love poetry from classical times through the
seventeenth century really do not seem to have a claim to public concern.) Some motifs are very simple but also so recurrent as to call for identification. One such motif soon recognized by all readers involves the embellishment of the woman. It may be something as much a part of her as a mole on
her bosom. It may be a beauty patch. It may be a rose worn
at her breast. No reader will be surprised that Herrick should
write "Upon the Roses in Julias bosome" or Carew "On a
Damaske rose sticking upon a Ladies breast," since they were
such accomplished observers of the happy life. But even
Habington, moral to the point of humorlessness, writes "To
Roses in the bosome of Castara." And Waller's "Go lovely
Rose" adapts the motif very finely indeed. From the adornment afforded by a flower, it is no long step to adornment by
milady's clothes. John Cleveland's "Fuscara; or The Bee Er-
rant" treats of her sleeve, "Where all delicious sweets are
hiv'd," and treats the woman in the time-honored way as herself the fairest flower. Sometimes the lady carries a fine object or looks at something fetching in the scene, and the very

model of such poems will be found in Waller's poem "Of a fair Lady playing with a Snake." When the lady is represented by things other than herself or as herself the fairest flower, her environment becomes an extension of her, as in some of Waller's poems on Sacharissa, or in some of Marvell's poems of country scene. Such poems are characteristically love compliments (or complaints, or both) and so this simple motif of embellishment may easily join such a scheme as the blazon.

Compliment may perhaps be thought of as an earlier stage of love, before intimacy provides its shorthand to courtesy. The question of the act of love itself provided another motif: "fruition." Should it be favored or not? It may seem cause for surprise that, at some points in their careers, Suckling and Rochester can be found arguing *against* fruition. One reaction to that may be to say that the motif must be very conventional indeed. It was, but its tradition enabled poets to extract both wit and titillation from it. From Ben Jonson to the Dryden-Tonson miscellanies one finds many variations on that motif attributed to Petronius: "Doing, a filthy pleasure is, and short"; therefore, "Let us together closely lie, and kisse."[37] Neither sexual delight nor sexual morality follows a straight line, and perhaps they cross more frequently than we admit. Rochester shows how "The Platonic Lady" could enjoy the maximum titillation and retain a technical virtue. She says,

> I'd give him liberty to toy
> And play with me, and count it joy.
> Our freedom should be full complete,
> And nothing wanting but the feat.
> Let's practice, then, and we shall prove
> These are the only sweets of love.   (19-24)

[37] *Poetae Latini Minores*, ci: "Foeda est in coitu et brevis voluptas." For Jonson, see poem xc in *Under-Wood*.

238

Rochester has managed to catch one of Jonson's accents and to reveal the "Platonic" hypocrisy. In his third sonnet ("Oh! for some honest Lovers ghost"), Suckling rejects unenjoyed love:

> T'have loved alone will not suffice,
> Unlesse we also have been wise,
> And have our Loves enjoy'd.   (12-14)

"Give me the Woman here" (l. 35) is the last dismissal of "Platonic" nonsense. But almost overleaf he writes "Against Fruition [I]," not that he wishes to impose a vow of chastity on the world, much less himself, but because " 'Tis expectation makes a blessing dear" (l. 23). Once again we may see how far the love poetry was concerned with the psychology of love, or with the motivations of passion, whether in observation of the lady, in compliment, or in consideration of the physical side of love.

Only the depraved, or at least the callous, however, can wholly dissociate the morality from the psychology of an act involving two people so intimately. Since our poets were not often callous, numerous of their poems treat related motifs of fidelity sought, protested, or affirmed; complaint over betrayal; or, and this shows better than anything else the psychological bent of the time, the argument showing that infidelity is either truth to nature or truth in some other guise. This motif, like so much else in the "schemes" and " motifs" of Cavalier poetry, is implicit earlier but emerges freely explicit in Suckling. He concludes "The guiltless Inconstant" with his last self-justification.

> So Love unto my heart did first prefer
> Her image, and there planted none but her;
> But since 'twas broke and martyr'd by her scorn,
> Many less faces in her place are born.   (27-30)

Here the explanation turns up an element of moral causation somewhat unexpected in Suckling, but the psychological vein we might expect is richer in the preceding lines. Waller has a similar (and better) poem, "To the mutable faire," but since he also answers Suckling's poem "Against Fruition,"[38] it would seem likely that Suckling started such fine games. The best poems of this kind are usually those involving a complex motivation for the woman as well as the man (e.g., Waller's "Cloris farwell I now must goe") or those combining a real human attachment with a sense of human vagaries (as in poems by Carew, Lovelace, Vaughan, and Cotton). Of course women have long had cause to require, and men to invent, explanations for male roving.

Two further motifs must bring to an end my consideration of love, for although either deserves a study to itself, I must move with greater haste. One of these enduring motifs identifies, or unites, or associates love and war. This motif is a very old Western, or pagan, conception and no doubt originally stemmed from rapine rather than from Bacon's theory of a soldier's compensating himself for hardship. The *Iliad* obviously joins the two elements, and "love and valor" provide lasting staples from Tasso and Aristo to Spenser, Davenant's *Gondibert,* and the heroic play. In such works, the association is often implied by plot rather than made explicit as a motif, so that perhaps we should credit Ovid for explicit use of the motive as a motif. From the clever first elegy of the *Amores,* he raises the possibility of identification of love and war. By *Amores,* I. ix, "Militat omnis amans," the identification is worked out in rich detail.[39]

---

[38] Waller, "In answer to Sir John Suckling's verses." Poems paired for or against something frequently appeared, sometimes by a single author.

[39] The standard concordance shows the frequency of this motif. For *miles* alone the frequency shown is: *Amores,* 15 times; *The Art of Love,* 7; *Heroides,* 7; *The Remedies of Love,* 1. See also in particular: *The Art of Love,* II. 233; and *The Remedies of Love,* l. 2.

Once again, Suckling provides what seems to be the first explicit Cavalier realization of what is today grown a cliché, "the battle of the sexes," and once again he employs a physicality that masks psychology. "A Soldier" presents the motif in terms of male sexual bravado. "Upon A. M." asks her to resign her fort only with struggle, since this soldier likes to fight at least such battles. On the other hand, in a poem often entitled "Loves Siege" (" 'Tis now since I sat down before") love's soldier shows himself willing to engage in the pleasing struggle only on equal terms, only as long as the woman acts on her own without calling in that ally, Honor.[40] When the Civil Wars became a reality, the motif must have seemed unusually piquant. It seems stretched to airy thinness in Lovelace's famous farewell to Lucasta as he set out for the wars:

> I could not love thee (Deare) so much,
> Lov'd I not Honour more.[41]

Perhaps at this point one must briefly recall the emblematic treatment of Venus and Mars, whether in painting or poetry, as the union of opposites from which Harmonia was born.[42] But I do not see such emblematic treatment in the *love* poetry of the Cavaliers. That *discordia concors* motif will be found in what the age called "heroic song," that is, in such poems as Denham's *Cooper's Hill*, Davenant's *Gondibert*, and Waller's *On St. James's Park*. It will also be found in a notable scene of *All for Love* and in that exciting hothouse, *Venice Preserved*. But in their "amorous song," the Cavaliers found their harmony in peace rather than in love's wars. Amorous battle provided much of the fun of passion, at least in poetic passion. But for lasting satisfaction, the Cavaliers sought steadfast peace.

---

[40] See also Suckling, "Upon my Lord Brohalls Wedding," ll. 29-35.
[41] Lovelace, "To Lucasta, Going to the Warres," ll. 11-12.
[42] See note 8, ch. iv, above.

Otway would also provide what Howell would term flex-animous examples of my last motif, the identification or association of love and death. In recent decades critics of seventeenth-century poetry have made much of the sexual overtones of "dying." They have made so much, in fact, that they have cheapened into a not so witty quibble what is in reality a complex range of associations and an abiding Western conception.[43] Returning to P. M., Gent. and his *Cimmerian Matron*, we read (p. 65):

> Notwithstanding Love be thus immortal, as being the proper affection of an immortal Soul, and devoted to an eternal Object, Good: yet can I not deny, but it is a kind of *Death*. For, who is ignorant that Lovers die as often as they kiss, or bid adieu: exhaling their Souls upon each others lips. Like *Apollo's* Priests possessed with the spirit of *Divination*, they are transported out of themselves; their life is a perpetual *Extasie*; they devest themselves of their own Souls, that they may be more happily fill'd with others. . . . Longing for the Elyzium of their Mistress breast, the only Paradise for Lovers Ghosts, they break the prison of their own, and anticipate the delivery of Death, and fly thither, as to the place of their eternal mansion.

The present state of our understanding of seventeenth-century thought and sensibility is not very advanced. It may be represented by the need, in discussing Cavalier love poetry, to cite a Restoration work to get modern critics off the sexual trail and into a fuller, more human view of love.

Death is longing, P. M. comes to tell us. In fact the only appropriate definition of "die" in the *Oxford English Dictionary* runs, in its entirety: "To suffer pains identified with those

---

[43] I have sketched the Western association of love and death as well as the Japanese of love and dream in "Japanese and Western Images of Courtly Love," *Yearbook of Comparative and General Literature*, xv (1966), 174-79.

of death; (often hyperbolical) to languish, pine away with passion; to be consumed with longing desire; *to die for*, to desire keenly or excessively." The definition fails to account for the sexual quibble periodically used, but it also fails to account for the full scale of P. M.'s idealism. We shall not err by steering on this side of bawdry. When P. M. speaks of lovers "Longing for the Elyzium of their Mistress breast," we recall that that paradise is the desired scene of poems as diverse as Carew's *Rapture* and Marvell's *Nymph complaining for the death of her Faun*. And the aspiration for love in realms "immortal," to use P. M.'s word, is at one with the mythopoeic tendency that we have seen in poetry from Jonson forward. But what is most important for a cluster of poetic motifs surrounding love-and-death is P. M.'s emphasizing "that Lovers die as often as they kiss, or bid adieu." In other words, lovers die from longing, they die from kissing or union, they die from parting, they die from absence, and they die from rejection. They are a dying lot, and one finds no surprise in Suckling's showing impatience.

> Why so pale and wan fond Lover?
> Prithee why so pale?
> Will, when looking well can't move her,
> Looking ill prevaile?
> Prithee, why so pale?[44]

Or again, in "Against Absence" (a repeated cause of lovers' deaths in the century), he shows that some men at least suffer more irrationally away from their mistresses than near them. His swashbuckling conclusion does not complete any very rigorous syllogism: therefore, the lover should give up the whole game and, by dining on sense, kill the killing love.

[44] Suckling, "Song" from *Aglaura*, iv. ii. For an unusual passage on the idea that a lover must be pale and pining, see Ovid, *The Art of Love*, 1. 729 ff.

Spare dyet is the cause Love lasts,
For Surfets sooner kill than Fasts.   (35-36)

There is no possibility, much less desirability, of searching Cavalier poetry like a critical coroner to record every death and its cause. Nor does there seem to be any need to reduce the love poetry to a rule and exercise of amorous dying. But the *ars amatoria* and the *ars moriendi* are good—and related —seventeenth-century concerns, as one quickly sees in Donne's poems, in Waller's "Go lovely Rose," and Marvell's *To his Coy Mistress*. Such concerns join most frequently not in poems on sexual union but in poems on parting, poems of valediction. Our training has led us to think first of seduction poems when the love poetry of the seventeenth century is concerned. But from Donne to Cotton, poets in all styles more often wrote valedictions. One of the elegies sometimes assigned to Donne and sometimes to Jonson is a valediction, "Since you must goe, and I must bid farewell." Although the poem's concern with time (and certain other considerations as well) lead me to think it more likely to be Jonson's than Donne's, the question of authorship concerns us less than the poem's traditional elements. There is that concern with wasting time and wasting love.

What fate is this to change mens dayes and houres,
  To shift their seasons, and destroy their powers!
Alas I ha' lost my heat, my blood, my prime,
  Winter is come a Quarter e're his Time,
My health will leave me; and when you depart,
  How shall I doe sweet Mistris for my heart?

(9-14)

He is in a dying way, especially since his heart is now hers. But of course he loves her; she *must* keep his heart:

And so I spare it. Come what can become
Of me, I'le softly tread unto my Tombe;

Or like a Ghost walke silent amongst men,
   Till I may see both it and you agen.[45]

Poems of lovers' parting are very rare with Herrick, but there can be few opportunities to part when there have been few to come together. Otherwise, Habington and Lovelace treat of parting and absence. So do Carew and Stanley, Waller and Suckling. And so do Davenant, Vaughan, and Cotton. Cotton has ten or more poems of parting, valediction, and absence. His "Song" ("Bring back My Comfort, and return") puts it all very simply: "That missing thee, I die" (l. 4). In "Taking leave of Chloris," he begins,

     She sighs as if she would restore
     The life she took away before.

And his "Ode Valedictory" declares that

     I go: but never to return:
     With such a killing Flame I burn,
     Not all th'enraged waves that beat
     My ships calk't ribs, can quench that heat.

                           (1-4)

The departures occasioned by the Civil Wars brought into being a number of poems in which the association of love and war and of love-parting and death found combination. Lovelace's poem, "To Lucasta, Going to the Warres," is the most famous and would demand quotation here if it had not been quoted from a few pages earlier. Davenant's "Song. The Souldier going to the Field,"[46] is characteristically rousing and eccentric. True, he can plead like any courtier:

     But first I'le chide thy cruel theft:
     Can I in War delight,

---

[45] The text quoted from here (ll. 19-22) is that of the *Poems of Ben Jonson*, ed. George Burke Johnston (London, 1954), p. 173, who has an extraordinarily packed note on authorship, pp. 333-34; modern opinion favors Jonson.

[46] Davenant, *The Works* (London, 1673), pp. 321-22.

> Who being of my heart bereft,
>   Can have no heart to fight?   (13-16)

But Davenant's true note sounds in the preceding stanza.

> For I must go where lazy Peace,
>   Will hide her drouzy head;
> And, for the sport of Kings, encrease
>   The number of the Dead.   (9-12)

Davenant was a friend of both Cottons, especially the elder. It was left to the younger, however, to join the two motifs for love, war and death, more fully—if also with less grim force. In "To Chloris. Ode" ("Farewell, My Sweet, until I come"), Cotton writes:

> Yet, when I rush into those *Arms,*
>   Where *Death,* and *Danger* do combine,
> I shall less subject be to harms,
>   Than to those killing *Eyes* of thine . . . .

> But, if I seem to fall in *War,*
>   *T'excuse the murder you commit,*
> Be to my Memory just so far,
>   As in thy Heart t'acknowledg it.

>   (9-12, 17-20)

    Sufficient reasons existed in the century to associate love and its partings with battle and with death. In the few remaining paragraphs of this chapter, I wish, however, to stand off from the schemes and the motifs and make a few general comments on Cavalier love poetry. The associations we have just been regarding belong to a very dominant tradition of the literature of love in the Western world, and in the seventeenth century they were not confined to lyric poetry. The tragical high passions of heroic love provide us with some of our best stories, and for our poets much the most popular was that of Dido and Aenas, whether the version by Virgil in the fourth *Aeneid* or by Ovid in the *Heroides*. (Denham almost seems the splendid poet of *Cooper's Hill* in his version.) Indeed, from the Middle Ages, and even in religious contexts (as P. M. had suggested in his way), death and love were at one. The *quia amore langueo* of the Spouse in the Song of Songs was taken as a type of Christ expiring on the cross out of love for mankind.[47] Freud merely glossed the images of the poets when he argued for the close association of Eros and Thanatos.[48] What such brief observations show is surely the extent to which Cavalier love poetry vibrated in tune with the Western chords of love.

One must not leave the impression that the Cavalier poet humorlessly called up the *anima mundi* as his familiar spirit whenever he wrote about love. The prominence given to Suckling in this chapter will confirm that the Cavaliers could be witty as well as traditional on the subject, but his example also shows that they were fundamentally psychological. Yet again he shows in some poems that they could also be humorous, which is a far more difficult tone to take on love. Entreaties to ladies had long before earned, and for good reason, the generic name of "complaints." But (to take another poet) Davenant's "Song" in Act III of *Newes from Plimouth* is a wonderful example of an entreaty that is fresh, and also of a kind that could be written in the England of the seventeenth and of no other century. Since it can make its own claims and must do so without being familiar to many readers, I offer it in its entirety without further introduction.

> O Thou that sleep'st like *Pigg* in Straw,
> Thou *Lady* dear, arise;

---

[47] Canticles, ii. 5 and v. 8. See Miner, *The Metaphysical Mode from Donne to Cowley* (Princeton, 1969), p. 240.

[48] Relevant passages by Freud are quoted in the article cited in n. 43, above.

Open (to keep the *Sun* in awe)
　Thy pretty pinking eyes:
And, having stretcht each Leg and Arme,
　Put on your cleane white Smock,
And then I pray, to keep you warme,
　A *Petticote* on *Dock*.
Arise, arise! Why should you sleep,
　When you have slept enough?
Long since, French Boyes cry'd Chimney-sweep,
　And Damsels Kitching-stuffe.
The Shops were open'd long before,
　And youngest Prentice goes
To lay at's [mistress'] Chamber-doore
　His Masters shining Shooes.
Arise, arise; your Breakfast stayes,
　Good Water-grewal warme,
Or Sugar-sops, which *Galen* sayes
　With Mace, will doe no harme.
Arise, Arise; when you are up,
　You'l find more to your cost,
For Mornings-draught in Candle-cup,
　Good Nutbrown-Ale, and Tost.[49]

Davenant's entreater seems to have taken us to London or Norwich, but he has many pleading brothers in poems by virtually all the Cavaliers. His tone must be compared with theirs entreating Corinna to go a-Maying, Lucasta to understand, Celia to yield, Chloris to give back the heart, and the Coy Mistress to act quickly. The colorings include many tones, humorous or witty, angry and profound. I cannot be-

---

[49] I have reversed roman and italic usage. Suckling's *Ballade. Upon a Wedding*, surely the finest mock epithalamion in the language, deserves mention for its even rarer combination of dignity with humor on a subject related to love. Its presence in numerous anthologies has led me to offer Davenant's less familiar but also worthy example.

lieve that Cavalier love poetry comprises the fullest and greatest canon on that most human of subjects. But along with new concepts of the religious experience, it seems to me to have assisted the seventeenth century in making its profoundly important step forward in understanding the human psyche. What had earlier required allegorizing to admit the validity of human love was an instinct to control one of the great human passions. One can admire allegory without much wishing to love allegorically. Suckling was too reductive in demanding, "Give me the Woman here," but he had a point. By looking into their real hearts without the neo-Platonic spectacles, the Cavalier poets could and did write. They also could and did return again and again to idealized as well as realistic versions of the psychology of love. By the middle of the century, some writers at least were again ready for neo-Platonic spectacles; but the main point to be made is that Cavalier love poetry, like all important poetic canons on that subject, combines real experience and numerous conventions for transforming the real into art. And certainly, whatever its status *sub specie aeternitatis*, Cavalier love poetry provides true versions of true experience in true art. We all feel, women as well as men, the attraction of "Come, my *Corinna*, come," and we remain forever ready for Amarantha, Delia, and Zelinda. Sometimes we shall smile when they come. And perhaps we shall grow sober when we observe Waller's grim smile at death when he tells his rose, "Then die. . . ." And when Jonson and Marvell urge ladies to "sport," we may agree that some meaning has been given us for hopes of happiness in the very jaws of devouring Time.

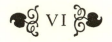

FRIENDSHIP

For friendship is nothing else than an accord in all things human, and divine, conjoined with mutual goodwill and affection, and I am inclined to think that, with the exception of wisdom, no better thing has been given to man by the immortal gods.

—Cicero, *De Amicitia*

THE UNEASINESS exhibited by seventeenth-century writers when they treated of love vanished when they turned to friendship. Here was an ideal without qualification. But it was also an ideal with numerous complexities, and one of those was that friendship tended to include love. To fulfill my warning toward the beginning of the last chapter, to show that there are problems in separating love and friendship into two chapters, I must quote Stanley's summary of Aristotle on these two lively topics.

Of *Love*, one kinde is of *Friendship*, another of *Conjunction*, and the third of *both*. The first is good, the second bad, the third mean.

Of friendship there are *foure* kinds: *Sodality, Affinity, Hospitality, Erotick*: whether that of *Beneficence* and that of *Admiration* be to be added to these, is doubtfull. The first is derived from *conversation*; the second from *nature*; the third from *cohabitation*; the fourth from *affection*; the fifth from *good-will* and last from some *facultie*.[1]

[1] Thomas Stanley, *The History of Philosophy*, 4 vols. (London, 1655-

Aristotle seems less Peripatetic than on the dead run in these two-sentence summaries of his views. But the direction he takes is plain: the best form of love is friendship, and friendship includes an *"Erotick"* sub-classification. One observes also an implicit ethics in such remarks, and such a moral strain surely appealed to all who wrote on friendship in the seventeenth century. To the Cavalier poets, who were so deeply conscious of their society and of time, friendship provided a way to follow the pattern of the good life here below in order to prepare for a good life above. One might earn a place in Elysium by being a friend, but to be a friend was to *do*. Cavalier poetry centers on the ethical rather than the ontological, and no readier example of the fact will be found than the poets' ideal preference of friendship over love.

i. Forms of Friendship

It will be recalled that in A *Rapture* Carew had urged his Celia to "flye with me to Loves Elizum" (2). My earlier remarks made no attempt to distinguish the grave efforts required to earn the blissful, transcendental state "in the noblest seates / Of those blest shades" above (23-24).

> The Gyant, Honour, that keepes cowards out,
> Is but a Masquer, and the servile rout

1662), II, sig. mmm 1ᵛ. He cogently sets forth the chief ideas of other schools on the head of friendship. Zeno's Stoicism is clear: *"A wise man only is a friend"* (II, sig. Nnnnn 1ᵛ). Similarly Pythagoras in a characteristic vein: *"He conceived the extremity (or end) of friendship, to be the making of one of two* [cited from Cicero, *De Officiis*, I. xvii. 56]. *Man ought to be one.* This sentence (saith *Clemens*) is mystick" (III, sig. Aa 2ʳ). This chapter takes a shorter but parallel path to that of David Latt in writing his dissertation, "The Progress of Friendship" (University of California, Los Angeles, 1971), and our sharing of ideas and references has been a source of delight to me.

Of baser subjects onely, bend in vaine
To the vast Idoll, whilst the nobler traine
Of valiant Lovers, daily sayle betweene
The huge Collosses legs . . .   (3-8)

There is something uncannily like Bunyan's lions, frightening
Christian, whose weak faith blinds him to the fact that they
are chained; and something, too, very like Cowley's "mon-
strous God which stood / In mid'st of th' Orchard" to
frighten away inquirers into knowledge.[2] To be able to remind one (even one as susceptible as myself) of both Bunyan
on Christian wayfaring and Cowley on the Royal Society is
to possess an ability to get at something important in seventeenth-century experience. Carew's wooing lover (after all,
by the end of the poem Celia has still not actually yielded),
like Bunyan, and like Cowley, makes strenuous efforts to
overcome obstacles and to get to a promised land. Doing,
effort, even struggle are required, and for love (as of course
for friendship, as we shall shortly see), doing required making
distinctions, choice.

Because Cavalier poets show little inclination to solve
present problems with ecstasy, their poetry has sometimes
been said to lack transcendence. "Ye kinder Gods," says
Suckling, "Give me the Woman here."[3] Or to recall Cowley
in his Metaphysical vein, the Cavaliers rejected "the strange
witchcraft of Anon!"[4] The trouble with "Platonick Love"
was in one sense that it was not love at all: to the Cavaliers
one had to possess at least a vision of the physical, the actual,
before any spiritual claims could be entered convincingly.
And yet the trouble with love was also that it was love; however "adamantine," in Burton's term, it was too physical, too

[2] Cowley, "To the Royal Society," st. 3.
[3] Suckling, "Sonnet III" ("Oh! for some honest Lovers ghost"), ll.
34-35. This is another of the Elysium poems of the time.
[4] Cowley, "Against Hope," l. 36.

transient, too subject to change. For such reasons, and for other reasons given in the last chapter, friendship claimed a much higher esteem. In a poem "To Mr. W. Hammond," Stanley addressed him, "Thou best of friendship, knowledge and of Art!" and he confessed his recent "Crime." Stanley had turned unfaithful to his friend by yielding to "female vanities," by falling in love. Fortunately, the lady showed a "stubborn heart" and that cured him of his "sin."

> That sin to friendship I away have thrown,
> My heart thou may'st without a rival own,
> While such as willingly themselves beguile,
> And sell away their freedoms for a smile,
> Blush to confesse our joyes as far above
> Their hopes, as friendship's longer liv'd then Love.[5]

One obvious reason why friendship could be thought superior to love is that it is "longer liv'd," a better remedy of time. Less obviously, it could also absorb the concepts, images, and emotional force of love. Stanley offers to exchange hearts, for example. But although lovers might call each other friends, they were never elevated by comparison to Orestes and Pylades.

The absorption of love motifs by poems of friendship can be clearly seen in Vaughan's *Poems* of 1646, which both begins and ends with an Elysium poem. It seems significant that whereas the existence or nonexistence of a real person behind his Amoret has led to some small debate, everyone feels that real friends lie behind the initials and references to friends. Amoret is not mentioned in the brief preface to his slim vol-

---

[5] Lines 29-34. For Stanley's relation with Hammond and other friends, see James M. Osborn, "Thomas Stanley's Lost 'Register of Friends,'" *Yale University Library Gazette* (1958), pp. 1-26. Osborn's is the *editio princeps*, with very helpful commentary. The poem is reprinted in Crump's edition of Stanley.

ume. But he does address his epistle "To all Ingenious Lovers of Poesie."

> Gentlemen,
>
> *To you alone, whose more refined* Spirits *out-wing these dull Times, and soare above the drudgerie of durty* Intelligence [that is, news], *have I made sacred these* Fancies: *I know the yeares, and what course entertainment they affoord* Poetry.

This is an outright Cavalier manifesto, Royalist in politics and preoccupied with the threat of the times to the good life. He continues:

> *If any shall question that* Courage *that durst send me abroad so late, and revell it thus in the* Dregs *of an Age, they have my silence: only,*
>
> Languescente seculo, liceat aegrotari;
>
> [When the whole world languishes, one may be permitted to be ill.]
>
> *My more calme* Ambition, *amidst the common noise, hath thus exposed me to the World: You have here a* Flame, *bright only in its own* Innocence, *that kindles nothing but a generous* Thought, *which though it may warme the* Bloud, *the fire at highest is but* Platonick, *and the* Commotion, *within these limits, excludes* Danger.

Now the readers addressed are "Gentlemen," and the merely ("but") "Platonick" element in the poetry is not that of love but of friendship. The Amoret poems are there for inspection, and someone more ingenious even than myself may find Platonism in them. But I think it more readily found in the higher strains of poems like the very first, "To my Ingenuous Friend, R. W." Vaughan begins, "When we are dead, . . ." The conditional is remarkable for setting off Cavalier from Metaphysical poetry in the canon of a poet who wrote both.

Donne or Herbert simply leapt the interval. Vaughan at this stage is Cavalier in seeing his vision in time. And during the Cavalier winter, the friends will need to endure a world of "noise," "sad tumults," and poverty before death.

> When all these Mulcts are paid, and I
> From thee, deare wit, must part and dye;
> Wee'le beg the world would be so kinde,
> To give's one grave, as wee'de one minde;
> There . . .
> Our soules shall meet, and thence will they
> (Freed from the tyranny of clay)
> With equall wings, and ancient love
> Into the Elysian fields remove.   (17-26)

Like the love poems, this speaks of Elysium, but with far greater assurance: the most persuasive lover might wonder if his lady would yield, but every friend knew that he and his friends would die. Carew's *Rapture* required the special terms of Elysium to exist, leading us to wonder whether this vision was possible in any sphere, at least on its own fancy terms. But Vaughan can claim R. W. as a real friend, and he can create a vision of "the Elysian fields" that both continues and extends friendship in ways credible to those who believe in a Christian afterlife. In those fields they will find "Great BEN" and "*Randolph*" and, walking on to "the drowsie fields / Of Lethe," they will see miserable lovers drinking the waters of forgetfulness, "That th' inconstant, cruell sex / Might not in death their spirits vex" (see ll. 37-50). So far does friendship appropriate the details usually associated with love poetry that the souls of Vaughan and his friend will, we are told, stay "on those flowry banks" (55) that were but "fables" (58) for love poets like Carew, but that are realized truth for the friends. Evidently any wise man would prefer friendship to love. To understand this theory, which must have been honored with practice on numerous occasions, we shall need to turn briefly to the doctrine of friendship.

The possible sources of Cavalier principles of friendship are numerous and impressive, going back, if not to David and Jonathan, at least to Empedocles and Heraclitus.[6] In the *Lysis* and the *Symposium*, Plato has Socrates develop friendship as a stage in attaining absolute Good. Of the ten books of the *Nicomachean Ethics*, Aristotle devotes two (VIII and IX) to friendship, treating it not merely as an impetus to virtue but also as a virtue in itself. His usual thoroughness does not fail him, and as Stanley's brick-wall redaction probably made clear, Aristotle codified many of the topics (e.g., equality) that were to recur in discussions of friendship over the centuries. Aristotle does, however, explain friendship in one set of terms, the political, that was available to Vaughan and is now lost to our ways of thinking.

> Moreover, friendship appears to be the bond of the state; and lawgivers seem to set more store by it than they do by justice, for to promote concord, which seems akin to friendship, is their chief aim, while faction, which is enmity, is what they are most anxious to banish.[7]

For Aristotle, then, concord resembles friendship and is preferred to justice; more than that, justice itself "is derived from consideration of relations suggested by friendship."[8] He also distinguished three meaningful systems of government—monarchy, aristocracy, and timocracy—each of which de-

---

[6] See Laurens J. Mills, *One Soul in Bodies Twain: Friendship in Tudor Literature and Stuart Drama* (Bloomington, 1937). Pages 1-10 give a useful survey of early literature on friendship. The scope of the book does not extend to our poets, but the ideas are very pertinent.

[7] Aristotle, *Nicomachean Ethics*, VIII. 1. 4 (Loeb ed., 1962), trans. H. Rackham. On concord (ὁμόνια) in such a context, see IX. vi.

[8] See Mills, *One Soul in Bodies Twain*, p. 7.

pended on friendship to preserve it from perversion.[9] It was perhaps because he seemed to have his eyes on the family, the state, and the people (the "real" world) that Aristotle and the Peripatetic school following him exercised greater influence on the doctrine of friendship in early modern times than did Plato, whose end was more philosophical in abstract ways. But the preference has also a simpler explanation: Plato's ideas lacked a full Roman expositor, whereas Cicero and Seneca propounded Peripatetic ideas at great length. Cicero of course also transmitted much of Plato, and for centuries little more of the Greek was known than what Cicero did convey. Plato's ideas were more fully absorbed and transmitted (along with some of Aristotle) by Plutarch, who had the ill judgment to write in Greek. It is Cicero and Seneca, above all Cicero, who were read in an age when, as we are particularly well equipped to understand, it was more common to mention the Greek worthies than to read them.

Before turning to the Latin moralists, however, we must again allow a moment, and not for the last time, to Quintus Horatius Flaccus. In matters of friendship, as in most matters in treating of the Cavaliers, one risks losing all proportion by omission of Horace. Even today, his *curiosa felicitas*, no matter how studied, is a relief from scholarly solemnity. (Even the baffled and shocked William Drummond continued to ply the English Horace, Ben Jonson, with amiable wine.) Horace provides a splendid model, then, and among the odes opening his collection, one finds that he has written one for Maecenas (i. i), a second for Augustus (i. ii), a third for Virgil (i. iii), and a fourth for Agrippa (i. iv). Horace's love of the country, of quiet, of wine, of poetry, and of a circle of

---

[9] Aristotle, *Nicomachean Ethics*, viii. i. 6: "the fairest harmony [ἀρμονίαν] springs from difference"; cited as a proverb to explain how different temperaments may lead to fast friendships—and demonstrating that Pythagoras and Plato did not have a monopoly on these tropes and saws.

cultivated friends possessed enormous appeal in the seventeenth century, and this Horatian ideal remains one that middle age has not had snatched away by the young. One might almost say that Horace exemplified the reasons why the Cavaliers could assent to the theories of friendship developed (or mostly rehearsed) by the classical moralists. And when we next meet Horace, it will be to discover that his epistles as well as his odes played a significant role in Cavalier poetry of friendship.

Although the Stoics emphasized self-sufficiency, Lucius Annaeus Seneca and other Roman Stoics revealed considerable capacity for friendship and affection. His essay, *De Constantia*,[10] which is so austere in ideas and so lively in manner, attempts to talk his Epicurean friend, Serenus, into accepting the Stoic doctrine of self-sufficiency. And had Horace lived to read Seneca's *De Vita Beata*, he would have attended to it, with some truancies, as a code for the *vita bona*. But after all, as we have seen in earlier chapters, the role of Seneca has been much exaggerated, at least relatively, and the central figure for defining what centuries of Western Europe have meant by friendship is Cicero.

Cicero sets the scene of *De Amicitia* at a time before he was born and gives most of what is said to Gaius Laelius, who credits most of his ideas to his close friend, the younger Scipio Africanus. Cicero himself says in his prefatory remarks to *his* great friend, Atticus, that he heard the substance of Scipio's ideas from Quintus Mucius Scaevola, his law teacher, who had heard it all from Laelius. What Cicero presents, then, is a doctrine of friendship which he traces, at least formally, back through three generations and which he attaches to Romans renowned for integrity and public service. Similarly,

---

[10] Seneca's "constancy" and that of his Renaissance imitators may be misunderstood today: *De Constantia Sapientis* refers to the unshakable mind of the Stoic sage.

there is a continuity into the future, since the lesser characters of the dialogue comprise a small group of friends who were to become well known after the fictional date of the dialogue. Such details may seem dry in the telling, but they show Cicero talking about friendship in a context of friends, of historical personages. Aristotle is interesting in the *Nicomachean Ethics*, but there is not so much there of Cicero's sense of personality and real life. On the other hand, by making his participants in the dialogue famous men, Cicero makes without stating it the Aristotelean point that friendship is the basis of the concord of the state. And he also implies what is implied by seventeenth-century poems on friendship: it is primarily the property of men, and it is entirely the property of those social classes that matter.

Actually, Cicero's *Of Friendship* is less the transmission of oral Roman tradition than a brilliant compendium, chiefly of Greek sources and especially from the pupils of Aristotle. Cicero's views on friendship turn out to be those of a man who has worthy friends, who has read the literature of friendship, and who has a mind capable of using what he has read to prove that he has friends. Written in those dangerous last years of his life, his dialogue is set in the context of the mysterious recent death of the Younger Scipio. One can easily see the appeal of such a situation to the men who tried to find their way over the slippery uncertainties of the Tudor and Stuart courts or the men who sought some source of strength during the vicissitudes of the Cavalier winter. One can also see why, given its situation, its tone, and its ideas, *De Amicitia* exercised such Stoic influence (Laelius was termed "the Wise" and is very like the ideal of Seneca nearly a century later). Perhaps no single work, not even Cicero's *De Officiis*, is so important for the transmission of Stoic ideas to sixteenth- and seventeenth-century England, albeit Stoicism with its rougher edges smoothed by other philosophies. And there are the great commonplaces. Only good men can be friends. Friendship is therefore necessarily unselfish. Above all friendship is natural, virtuous, and conducive to that rational happiness which is the end of life. It enables us to bear the sufferings that we must endure. And, what is perhaps the commonplace to re-echo most often, a friend is another self (*De Amicitia*, xxi. 80).

The values of friendship are sufficiently simple or basic as not to require extended discussion. I have tried to suggest that one only need know Cicero's dialogue (and also that one must know it) for one's background to be sufficient, if not complete. One can demonstrate fairly easily the essential character of friendship by the numerous works published on the subject. But yet better proof lies in the fact that it is the one topic on which there is no disagreement in the century. The religious merely added that Jesus is the true friend. Even Millamant, in 1700, recognizes it as a natural thing that Mirabel will choose his own friends and that they will occupy a sphere of his life other than that centered on her. Friendship was not a subject on which anyone in the century had to make up his mind, and for that reason alone it must have trebled in popularity each decade. For that very reason, however, friendship did present the poets with the problem of demonstrating the accepted virtues of friendship in usable poetic forms.

In this, as in so much else, those who explored the realm of friendship found that most of its prominent features bore the tracks of Ben Jonson. For some time he used the epigram as a form, achieving notable success in creating the sense of the virtue of his friends (and himself) and also the conviction of his friendship with William Camden, John Donne (two poems), Sir John Roe (three poems), Robert Cecil, Earl of Salisbury (two poems), Francis Beaumont, Lucy, Countess of Bedford, and others. His pindaric ode is not only, it will be recalled, an elegy on the *Immortall Memorie* of Sir Henry Morison but also on the *Friendship* of Morison and Sir Lucius Cary, with Ben Jonson himself playing no small role. On reflection, Jonson's practice leads us to understand how important friendship is to most of the finest funeral elegies, and after recalling *Lycidas*, we remember how often the pastoral elegy extols or is founded on friendship. Still other forms were employed. The commendatory poem for prefixing to a volume of works by a living or dead author was one such additional possibility, and again we see that Jonson seized it and made it the basis of some of his finest poetry, as the commendatory poem for Selden's *Titles of Honour* or for the first Shakespeare folio. In the drama, from Shakespeare to Otway, the "duel of friends" became a major set-piece.[11] But it remains here to attempt description of what has proved to be one of the more indescribable literary forms, the one that dominated the field for poems of friendship. It is what some people have termed the verse epistle, *metrica epistola*.

Like the dialogue, the letter is a most appropriate form for representing intercourse between friends. For his "Familiar Letters," *Epistolae Ho-Elianae*, James Howell used a motto that may be translated, "As the key unlocks the door, so the letter unlocks the breast."[12] The natural English term is of course "letter," and until early in the seventeenth century "epistle" was usually restricted to the apostolic epistles in the New Testament.[13] (Curiously, I have missed any mention of

---

[11] See Mills, *One Soul in Bodies Twain*, ch. v. The "duel" involves an argument that serves, upon resolution, to affirm the relationship. I may add that the false friend is another sub-topic in the lore of friendship.

[12] "*Ut* clavis portam, *sic pandit* Epistola pectus." Howell's subtitle is *Familiar Letters Domestic and Foreign; Divided into Four Books: Partly Historical, Political, Philosophical. Upon Emergent Occasions.*

[13] See the *Oxford English Dictionary*, s. v. "Epistle." This latinate

such a New Testament epistle as Philemon in seventeenth-century discussions of friendship or of the letter.) The usage of "epistle" for the verse letter also appears at the beginning of the century. In the dedication of his first volume of *Epistles* (1608), Joseph Hall speaks of "a new fashion of discourse, by Epistles; new to our language."[14] What is involved is the gradual evolution in English practice and awareness of the natural letter into an art form. Once the breakthrough had been made, success came quickly, as the instance of James Howell shows: the *Epistolae Ho-Elianae* was frequently reprinted, even in that century of great letter-writers, the eighteenth. More than that, he knew the value which the age attached to friendship and the relation of letters to the value.

> Love is the Life of Friendship, *Letters* are
> The Life of Love, the Loadstones that by rare
> Attraction makes Souls meet . . .[15]

That this is a commonplace can be judged by comparing it with Burton's remarks on love quoted at the head of the last chapter. And, as Howell goes on to say, letters may be in poetry or prose.

> *Letters* may more than *History* inclose
> The choicest Learning both for Verse and Prose.

So he goes on: "The bashful Lover" (i.e., friend) becomes eloquent with his quill; letters are "Ideas" of the rational soul. At the same time, if letters are so important for what

---

word had the advantage over "letter" in that it was capable of derived forms such as "epistolary" and "epistolize." These attained full currency in the first third or half of the seventeenth century.

[14] *Ibid.*

[15] Howell, *Epistolae Ho-Elianae*, 11th edition (London, 1754), p. 13: "To the Knowing Reader touching Familiar Letters."

they reveal of the individual, and if they are so essential to friendship and love, they also unite the state, as does friendship in Aristotle's view. In fact, more than half his versified address to the reader (that is, the longer middle paragraphs, from the third through the seventh) treat of just this larger topic.

> *Letters* like *Gordian* Knots, do Nations tie,
> Else all Commerce, and Love, 'twixt Men would die.

When, therefore, Lovelace addressed the elder Cotton as the "best of *Men* and *Friends*" during the time of Royalist defeat, he was writing more than an Anacreontic poem. He was writing an adapted version of the epistle. And he was doing yet more: by affirming a true friendship he was affirming a true polity for the state, what was natural and just (in the Royalist view), even if the new, unjust laws had triumphed over the old true state. I do not know how to emphasize this neo-Aristotelean notion strongly enough. It is implicit in Cicero, but all one can do finally is quote, and here is Howell again: "Friendship is the great Chain of human Society, and Intercourse of Letters is one of the chiefest Links of that Chain."[16] Since he was a friend of Ben Jonson, Sir Kenelm Digby, Endymion Porter, and the Wroths, he was able to generalize from the ideals of some of the central figures among the early Cavaliers.

We think that we know well enough what friendship is, and we possess similar assurance over the natural letter. For the seventeenth century (the eighteenth regulated such things better), it is far more difficult to set forth the criteria by which to distinguish an "art" letter from its "natural" parent. And exasperation, at least of any high hopes one might have had, faces the student seeking the criteria of the

---

[16] *Ibid.*, i. 2. xviii (pp. 106-107), 3 August 1622; the text followed is in italics.

verse epistle.[17] Howell, who styles himself with more justice "the worst of Poets" than Charles I "the best of Princes," not even Howell proves immune from error. But he at least reveals a self-consciousness, or an explicit awareness of form, that one searches for in vain among the literary critics of the century. And yet even his poem "To the Knowing Reader touching Familiar Letters" raises problems. Is it a verse letter? How can one write a familiar verse letter to the knowing reader when one does not know who one's readers are? Or again, natural letters have conventional openings and closes and grow from a temporal situation usually important enough to require dating. Howell's verses, like Jonson's *Epistle to Master John Selden*, do not observe such conventions. Neither did Horace in his *Epistulae*—so why, it may be asked, why complicate what is plain? Unfortunately, it is not plain to the classicists how one can distinguish between a Horatian epistle and a Horatian satire, except that some are called by the one name and some by the other. On the other hand, of his odes, which are not written in the same prosody, one-third are addressed to individuals. Is it only the prosody that makes the difference: must a Latin verse epistle be written in dactylic hexameter? Catullus and Ovid prove awkward here in using the elegiac distich to that end. And although Ovid in the *Heroides* uses the heroic measure, he explores the wholly different world of love so well that most commentators have thrown up their hands and arbitrarily decreed that there is a didactic Horatian epistle and an amatory Ovidian epistle. Neither category much assists our understanding of poems of address between friends.

17 In what follows I draw on my essay, "Dryden's *Eikon Basilike*: To *Sir Godfrey Kneller*," in *Seventeenth Century Imagery* (Berkeley and Los Angeles, 1971), edited by me; numerous sources and certain significant modern studies are treated there. The *dicta* of Renaissance classicists and the studies of modern classicists are also examined.

When fine distinctions do not seem to work, we must turn to grosser characterizations. Let us assume that there may be some transformation of natural letters into various forms of poetry and prose, and then let us see what properties, if any, these various forms show. What a very large number of seventeenth-century poems have in common is a poetic address from the poet to the addressee. I do not mean that Donne's poem "The Flea" is a Horatian epistle. There the lady is inside the poem, and it is ten to one that she never had an address, since she never had a name. Poetry of address requires the social mode spoken of in the first chapter (or the public emerging so strongly after the Restoration). The poet who addresses and the person who is addressed meet on terms of social relation: friend and friend, father and sons (as in the Tribe of Ben), pupil and teacher (Jonson and Camden), subject and prince, and so on. There is a sense as well of other people about, other friends, and enemies, too. We seldom get ecstasy in Cavalier poetry for this very reason. Ecstasy, whether amatory or religious belongs to a "fine and private place." Friends may share such intimate scenes and may testify to their love for each other. But they may also stroll arm-in-arm in the country or through the town, with their concord or love being so far from an offense to others that they are (in Aristotle's word) a microcosm of the state.

Poetry of address also requires demonstration of common interest between the poet and the addressee. Their interests must coincide, or at least the nature of their social relation must have brought them together for social purposes. This is how friendship becomes so ready a subject; this is why poetry of address is so often the form used to treat friendship. One cannot claim that there is anything particularly "seventeenth-century" (or "late Renaissance") about this, as Chaucer's wonderful addresses to his friends, to his copyist, and even to his purse all show. What is particularly "seventeenth-century" is the high proportion of important poems sharing the characterics of poetry of address. So far we have seen only two or three characteristics of verse address, and in fact I can add only one more. Isolatable characteristics common to most examples are few, in part because we are dealing more with a transformation of a natural form (the actual letter) itself, and partly because there just are not very many formal energies common to poems in our category. Nonetheless the poetry of address exists; it somewhat resembles political parties or the eighteenth-century essay. Hard to define all right, but evident as sunshine. I can add only one further characteristic: in the poem of address the poet addresses himself in a double sense, that is, both to a person and a topic. Such matters come together in most of the best poems of address, and the poet makes his topic important not only between the speaking poet (or "persona") and his addressee but also between the poet and his readers. Not all poems touching on friendship are poems of address, nor are all poems of address concerned with friendship. Some pastoral dialogues treat friendship, and some poems of address seek out enemies rather than friends. Such distinctions will not burden any reader, however, and we may now turn to a very different kind of discrimination.

ii. The Few Good Friends before the Wars

Poems treating friendship discover varying significance in the relation as the century advances. For utility's sake, we can distinguish three periods: before the wars, the Cavalier winter of oppression, and rural friendship while awaiting the King's return. Such division implies no close dating, but some such sequence does seem to hold for developments in ideas about friendship as much as it does for the history of the period. The given emphasis reflects the experience of three different generations, or of three stages in the life of one who lived from 1580 to 1660. What above all distinguishes the poetry of friendship before the wars is the Jonsonian moral stance. He can look into himself and say that he is true ("alwayes one"). He can look into those few good men about him and claim them friends. He can seal them into his Tribe, or he can counsel a friend to fly the country and seek honorable death in war rather than ignominious corruption at home. All this sounds terribly severe, and some of it is. But there is a lot of good humor, too, especially when Jonson addresses Vulcan or himself. And his most dutiful Sons enjoyed winking at him. To show as much I may take up Randolph where I left him many pages back in "A gratulatory to Mr. Ben. Johnson for his adopting of him to be his Son."

> the whole quire
> Of Poets are by thy Adoption, all
> My uncles; thou hast given me the pow'r to call
> *Phoebus* himselfe my grandsire, by this graunt
> Each sister of the nine is made my Aunt.   (14-18)

Then, turning to the old commonplace that true nobility stems from virtue rather than blood, Randolph gives his humor a somewhat different coloring.

> Go you that reckon from a large descent
> Your lineall Honours, and are well content
> To glory in the age of your great name,
> Though on a Herralds faith you build the same:
> I do not envy you, nor thinke you blest
> Though you may beare a Gorgon on your Crest
> By direct line from *Perseus*; I will boast
> No farther then my Father; that's the most

I can, or should be proud of; and I were
Unworthy of his adoption, if that here
I should be dully modest; boast I must
Being sonne of his Adoption, not his lust.   (19-30)

I am not sure that Randolph ever succeeded as a poet when
he was *fully* serious, but in the remainder of the poem two
topics are raised that join him to his great addressee, Jonson.
The first is literary: surely a son may steal Promethean fire
from his father. (The allusion is a very nice one in the context, that such an old rebellion should serve the ends of filial
piety.) The second is personal but related: he would like to
borrow some heat from Phoebus to cure Jonson's palsy. And
Phoebus had better give it,

                    else I will complaine
He has no skill in hearbs; Poets in vaine
Make him the God of Physicke; 'twere his praise
To make thee as immortall as thy Baies;
As his own *Daphne*; 'twere a shame to see
The God, not love his Preist, more then his Tree.
                        (57-62)

No wonder Randolph charmed Jonson and made a whole
generation think (on no better evidence than this poem,
surely) that he was heir as well as son to Ben Jonson.

Jonson's capacity to take criticism as well as praise from his
friends and their giving of both, provide important measures
of his moral stature. In theories of friendship then, long before then, and even now, a friend is considered a person who
will speak the truth to us. In practice, we often fear that
candor will wound the self-esteem or dash the spirits of a
friend, and we are apt to think, when a friend criticizes us,
that even he does not understand us, or that even he has
turned enemy. Jonson encouraged a remarkable candor from

his friends, as we have seen with Howell. There also occurred the fuss over *The New Inne*. When this play (one of his dotages, as Dryden put it) failed on the stage, Jonson wrote a poem vindicating the published version and blasting the depraved tastes of the age. Thomas Carew was willing to concede the faults of the age, but in "To Ben. Johnson" he put it as straight as anyone could put it to a friend who was capable of receiving criticism.

> Tis true (deare *Ben:*) thy just chastizing hand
> Hath fixt upon the sotted Age a brand
> To their swolne pride, and empty scribbling due,
> It can nor judge, nor write, and yet tis true
> Thy commique Muse from the exalted line
> Toucht by thy *Alchymist,* doth since decline.
>
> (1-6)

Carew concludes mingling perfect manners with his criticism.

> The wiser world doth greater Thee confesse
> Then all men else, then Thy self onely lesse.
>
> (49-50)

What Carew was saying was that Jonson's response revealed an "ytch of praise" (26) that was unnecessary. Whether Jonson was displeased by such frankness is a question we cannot now answer very well; one doubts that he can have been very pleased by the figure Carew showed him to have cut. Yet, as he read his friend's verses another time, he must have observed how far Carew had attempted to follow Jonson's own moral style in poetic address. Imitation, they say, is the sincerest form of flattery and is especially welcome when, as in Carew's case, the style fails to offer any rivalry to the original. Howell, we recall, said very nearly the same thing about Jonson's late plays.[18] But for the next six months

---

[18] Howell, *Epistolae Ho-Elianae,* I. 5. xvi (p. 214).

he sought to execute Jonson's commission to find a copy of "Dr. *Davis's Welsh* Grammar," which at last he offered to his Father "for a New-year's-gift," enclosing with it a letter and a poem.

Jonson was obviously admired for something including his plays, his poems, his learning, his powers of speech, and his formidable will.[19] Such matters were important, and each was capable of human imperfection on a scale commensurate with Jonson's achievement. But for all his faults, which were enough to provide Juvenal with another satire on the poet as dictator, Jonson possessed something so fine that it transcended his faults and inspired his other virtues. That something is humanity, a fearless manliness, and a candid integrity. His mettle was right, sound, and true, and these qualities still shine in his poetry, affording those with experience of this world and themselves a better light to judge the darkness we live in than the flashy skyrockets we worshipped in our youth. In his "Epistle To a Friend," a title made to order for this chapter, Jonson apologizes for not having performed what he had promised at the time promised, developing for friendship a relation taken from what we would term economics.

> They are not, Sir, worst Owers, that doe pay
> Debts when they can: good men may breake their day.
>
> (1-2)

(Comparison of this and the next quotation with the last from Carew will suggest how far he emulated Jonson's style

[19] Jonson's many-sidedness is brought out better than space allows me to try doing by George Burke Johnston, *Ben Jonson: Poet* (New York, 1945). Johnston is of course also editor of the convenient Muses' Library edition of Jonson's poems. I must also commend for the virtues proved by my students and myself, the edition by William B. Hunter, Jr., *The Complete Poetry of Ben Jonson* (New York, 1963), which is the most usable edition.

and show his degree of success in doing so.) Jonson's interest lies in the spirit of a pledge, not merely the legal bond, or "Band."

> for he that takes
>   Simply my Band, his trust in me forsakes,
> And lookes unto the forfeit. If you be
>   Now so much a friend, as you would trust in me,
> [Venture] a longer time, and willingly:
>   All is not barren land, doth fallow lie.
> Some grounds are made the richer, for the Rest;
>   And I will bring a Crop, if not the best.   (11-18)

In the years before the wars, Jonson and his friends culti-vated those fields of friendship; theirs was a plentiful harvest. But we should not become sentimental, nor should the days before the wars be regarded as a time when the sun always shone. Quite the contrary: the friends were the happy and good few, but the most were a very sorry lot indeed. It will be recalled how grim a picture Jonson draws in another poem sharing the title of the last: *An Epistle to a Friend, to Per-swade Him to the Warres*. Like many of the epistles of Horace, this address seems to have become a satire. In other words, the vision of evil around one does not merely function as a rhetorical device to set off praise. Jonson gave expression to his vision of evil because that was what he saw in the "number" of the bad. No wonder the good were to be prized.

> 'Tis growne almost a danger to speake true
>   Of any good minde, now: There are so few.
> The bad, by number, are so fortified,
>   As what th'have lost t'expect, they dare deride.[20]

It will be recalled that this poem to Lady Aubigny includes

[20] Jonson, *Epistle. To Katherine, Lady Aubigny*, ll. 1-4.

that essentially Jonsonian line, "For he, that once is good, is ever great" (l. 52).

But why write poems about these matters? One answer will be found in the confession Jonson makes that he has sometimes praised people more than they deserve. As he writes in his *Epistle to Master John Selden*, his aim had been to improve those he praised.

> Though I confesse (as every Muse hath err'd,
> And mine not least) I have too oft preferr'd
> Men, past their termes, and prais'd some names too much,
> But twas with purpose to have made them such.
>
> (19-22)

Behind this lies not only Jonson's frank awareness of himself, but also the old view that poetry had moral effect. In the following lines, Jonson seems to one reader to get closer to the truth about what literature is like.

> Since being deceiv'd, I turne a sharper eye
> Upon my selfe, and aske to whom? and why?
> And what I write? And vexe it many dayes
> Before men get a verse: much lesse a Praise;
> So that my Reader is assur'd, I now
> Meane what I speake: and still will keepe that Vow.
>
> (23-28)

I shall say again that we can regard this as a means of setting off praise, and a bit later I shall return to the issue. For now, let it be said that of course it sets off the praise of Selden, but the contrast between good and bad is not less valid because a poem deals with both elements. And let us look therefore more carefully at Jonson acknowledging his faults and asking *why* he writes. This is the central question at the moment, because if we answer it we shall understand the bond of friendship that drew Jonson to the best people of his age.

A simpler poem, "To Mary Lady Wroth" (*Epigrammes*, ciii), may reveal things that are less explicit in more complex poems. To this niece of Sir Philip Sidney, Jonson dedicated *The Alchemist* and wrote three poems, as well as one to her husband. Jonson contrasts what the Muses of other poets may lead them to say with what his Muse will so simply advance in her praise.

> Forgive me then, if mine but say you are
>   A *Sydney*: but in that extend as farre
> As lowdest praisers, who perhaps would find
>   For every part a character assign'd.
> My praise is plaine, and where so ere profest,
>   Becomes none more then you, who need it least.
>   (9-14)

The evidence stands before us; the problem is clear: the good do not *need* praise and the bad do not mend when they are blamed. Why indeed write? Pope faced the same question in his *Epistle to Dr. Arbuthnot,* an *apologia* modeled, like some of Jonson's poems, on the Horatian epistle. And as Pope confesses to his friend (ll. 83-108) the bad continue bad for all his satire. I think that Jonson gives the real answer to our question in another epigram (cii), "To William Earle of Pembroke."

> I doe but name thee *Pembroke*, and I find
>   It is an *Epigramme*, on all man-kind;
> Against the bad, but of, and to the good.   (1-3)

One writes, then, to discriminate bad from good, not to change the bad. One writes also to show that one can discriminate between vice and virtue and that one will affirm the good.

> Nor could the age have mist thee, in this strife
>   Of vice, and vertue; wherein all great life
> Almost, is exercis'd: and scarse one knowes,
>   To which, yet, of the sides himselfe he owes.
> They follow vertue, for reward, to day;
>   To morrow vice, if shee give better pay:
> And are so good, and bad, just at a price,
>   As nothing else discernes the vertue' or the vice.
> But thou . . .   (5-13)

One writes because one simply must speak the moral truth (if there is any other kind in this context), and that is why Jonson's friends could address him as they did.

There are other considerations. The good man triumphs over time.

> But thou, whose noblesse keeps one stature still,
>   And one true posture, though besieg'd with ill
> Of what ambition, faction, pride can raise.   (14-16)

Celebration of the good man or woman, one's friend, also triumphs over time, because art itself confers immortality, at least when it affirms truth; for as Jonson wrote Katherine, Lady Aubigny,

> as long yeeres doe passe,
> *Madame*, be bold to use this truest glasse:
> Wherein, your forme, you still the same shall finde;
> Because nor it can change, nor such a minde.
>   (121-24)

Given the mimetic theory of art, one writes to show the truth, and in mirroring it, art's looking glass itself takes on perpetual truth. In a sense, then, although art changes nothing, making no good man a Tom Coryat or no bad a Sir Philip Sydney overnight, it makes the good and bad clear in their properties, and it makes them last in their essential states. And so, to return to friendship, the discrimination of bad from good and the celebration of good is central to poetry of friendship because friendship itself, rightly understood, is truth active in human relations. The poetry of friendship sustains and continues the little society of the good few, and it demonstrates as well powers of mind and feeling. In addressing John Selden, Jonson praises another friend of Selden's, Edward Hayward, as just such a knowing one who

> will not only love
> Embrace, and cherish; but he can approve
> And estimate thy Paines; as having wrought
> In the same Mines of knowledge; and thence brought
> Humanitie enough to be a friend.   (73-77)

O Rare Ben Jonson! The inscription in Westminster Abbey seems exact, for Jonson is rare as an extraordinary individual and as one who created a wide variety of poetic means to sustain a vision of friendship. That vision retains value for its merger of ethical realities with mythic capacities in a language of rare economy and manliness. Perhaps of all the tributes in *Jonsonus Virbius* one of the most moving is also one of the simplest. Sir John Beaumont ends his poem with just the right claims.

> Could I have spoken in his [Ben's] language too,
> I had not said so much, as now I doe,
> To whose cleare memory, I this tribute send
> Who Dead's my wonder, Living was my Friend.[21]

Can the matter admit any doubt?

Herrick, another friend of Jonson's nowadays so often appears like Ovid, as the wooer of Corinna, that his poems on

---

21 *Ben Jonson*, ed. C. H. Herford, Percy and Evelyn Simpson, 11 vols. (Oxford, 1925-1952), XI, 439.

friendship grow too little known. At times, however, he is more successful than Carew in catching Jonson's cadence and thought. *A Panegerick to Sir Lewis Pemberton*, it is true, begins with more of Herrick.

> Till I shall come again, let this suffice,
>   I send my salt, my sacrifice
> To Thee, thy Lady, younglings, and as farre
>   As to thy *Genius* and thy *Larre*. (1-4)

But we find that the strains of the Jonsonian music grow stronger.

> Thou do'st redeeme those times; and what was lost
>   Of antient honesty, may boast
> It keeps a growth in thee; and so will runne
>   A course in thy Fames-pledge, *thy Sonne*.
>
> (41-44)

If Jonson has a lineal successor, Herrick is surely the legitimate one. But he is also his own man. The ritualistic paganism of his poem—with its "salt," "*Genius*," and "*Larre*" reveal his hand at once. His verse form (couplets made up of a pentameter and tetrameter) obviously takes the elegiac distich as its model, and with the elegy Herrick combines panegyric, country-house poetry, and poetic address of the kind we have been following.[22] It must not be pretended, however, that the address is the most important element, or even friendship; rather it is the Jonsonian vision of ethical

---

[22] In his excellent edition of Herrick, *The Complete Poetry* (New York, 1968), J. Max Patrick compares to Herrick's poem the following: Joseph Hall, *Virgidemiarum*, v. 2; Martial, III. 58; Jonson, *To Penshurst*; Carew, *To Saxham*; and Marvell, *Upon Appleton House*. Patrick would of course recognize that many other poems fall in this general category, and it would be difficult to say at what stage the seventeenth-century conception alters into something else, whether with Pope's *Windsor Forest* or some later work.

truth that made poetry on friendship a social and artistic ritual. Herrick deserves quoting at length on the architecture of virtue, speaking of

> What *Genii* support thy roofe,
> *Goodnes* and *Greatnes*; not the oaken Piles;
> *For these, and marbles have their whiles*
> *To last, but not their ever*: Vertues Hand
> It is, which builds, 'gainst Fate to stand.
> Such is thy house, whose firme foundations trust
> Is more in thee, then in her dust,
> Or depth, these last may yeeld, and yearly shrinke,
> When what is strongly built, no chinke
> Or yawning rupture can the same devoure,
> But fixt it stands, by her own power,
> And well-laid bottome, on the iron and rock,
> Which tyres, and counter-stands the shock,
> And *Ramme* of time and by vexation growes
> The stronger: *Vertue dies when foes*
> *Are wanting to her exercise, but great*
> *And large she spreads by dust, and sweat.*   (98-114)

Here is Cavalier poetry in microcosm: the social mode, the good life, time and its remedies, and order and disorder, all enter naturally; and although there is nothing of love except within the family of the house, there is that air of candid friendship ("Till I shall come again, let this suffice") that makes Herrick's idea of a "Panegyrick" different from Donne's or Dryden's. No one would dispute the interest of the poem in its own right, either. But I also fancy that one feeling we have while reading it is that no one else has imitated Jonson so well. I hope that it will not be thought a criticism of Herrick to say that his poem does lack Jonson's powers of concentration of much observation into little space, that distillation of life into poetic truth. After all, no other

English poet excels Jonson in this. All these things would be shown by reading *To Penshurst* again.

Herrick does, however, have a yet finer poem in "elegiac" couplets, *A Country Life: To his Brother, Master Thomas Herrick*. Suppose I had been set these lines to identify for authorship:

> By studying to know vertue; and to aime
> More at her nature, then her name:
> The last is but the least; the first doth tell
> Wayes lesse to live, then to live well.   (7-10)

I should have failed, marking them "Jonson." Let anyone else put his hand on his heart and declare that he would have said, "Herrick." But the poem as a whole is more like the rector in Devon than the sage of London, in that it contents itself with being a Herrick interested in those telling little things that make up the world rather than a Jonson taking the great world apart to find its principles. All of us can feel we have our poetic geography straight with the essentially Herrickian hearth scene.

> Yet can thy humble roofe mantaine a Quire
> Of singing Crickits by thy fire:
> And the brisk Mouse may feast her selfe with crums,
> Till that the green-ey'd Kitling comes.   (121-24)

To have integrated such diverse strains into a single poem is no middling feat. (For example, the Country Mouse is altogether Horatian in origin and behavior, finding her own version of a country estate when danger threatens.) Except in *Corinna's going a Maying*, Herrick's touch was never so sure and so firm at once; and although *that* poem is more "sincere" in the seventeenth-century sense of being purer, unmixed, and although it is finally more urgent, the poem to his brother and his wife proves to be more adult, more instinct with what life is all about.

Perhaps it seems unclear why this or the poem to Pemberton are included in a chapter on friendship, or why, if they treat such heads, they are poems of praise. The answer proper to the first doubt will be found in the wide range of associations to "friendship" from Aristotle onward. In highly stratified societies it would inevitably happen that a younger or socially inferior person might become a valued friend of an older or superior one. The relation might even be that of patron and patronized as much as that of friend and friend. Shakespeare's relation with Southampton and Horace's with Maecenas were such. Or, to take an opposite pole of intimacy, lovers were also considered to be friends, and friendship (as we have seen) the highest form of love bar charity, with which it was even identified at times. But perhaps a simple illustration will serve best. Each of the three parts of William Habington's *Castara* (in its complete form) has a prose character: the first "A Mistris," and the third "A Holy Man." But the second has two such characters, "A Wife" and "A Friend"—the two personalities fit into a single conception of life. The explanation for the presence of praise requires only pointing to the obvious fact that it offers one means of telling your friend you like him, of saying that he (or she) is your friend. Beyond so simple an explanation there lies a moral imperative, for just as Jonson bequeathed a tradition of satiric attack on the bad, so he handed down a tradition of the praise of the good. In Herrick's poem itself, we also sense how natural it is, without worrying a whit over literary traditions, to praise those one loves. In another sense, praise is required because of the urgent sense that life is short and vexed with trouble. Herrick's "brisk Mouse" scurrying off "to her Cabbin" turns out not to be merely a pretty detail but an emblem

279

of man's estate: "And thus thy little-well-kept-stock doth prove . . ." (l. 127).

The code of Herrick's country life centers on the Horatian mean (cf. *Odes,* I. xxvii): "still conning o'r this Theame, / To shun the first, and last extreame" (131-32). Certainly life there is not thought of in merely moral terms: we see the familiar paradisal image for happy love: "Thus let thy Rurall Sanctuary be / *Elizium* to thy wife and thee" (137-38). I have deliberately refrained from quoting more than brief passages from this splendid poem, in order to stress the variety of subject and tone to be found in it. And yet, one may defy any reader to do two things: to read the poem without thinking that its view of life holds together; or to have been prepared for the charge at the end of the poem that the husband and wife should not only be happy together, or be good together, but *die* together. Herrick sounds ripest in poems like these, and if they do not dispense with the necessity for our knowing *Corrina's going a Maying,* that surely is a mercy of poetry and of life.

The poems of complimentary address by Jonson and Herrick treat the good life in terms of those good men and women who are their friends. Herrick seldom raises Jonson's concern with real and present danger (although one kind of major exception will shortly appear). As we might have expected, the good life sought by Herrick sometimes turns out to be the *vita beata* rather than the *vita bona.* The poem whose title is complete, even if the poem is not—"The Country life, to the honoured Master Endimion Porter, Groome of the Bed-Chamber to His Majesty"—very much emphasizes the happy life. Perhaps the poem proves conventional. But I frankly find something real and lively in this poem, an evocation of that greenest and loveliest of English counties, Herrickshire.

These seen, thou go'st to view thy flocks
Of sheep, (safe from the Wolfe and Fox)
And find'st their bellies there as full
Of short sweet grasse, as backs with wool.
And leav'st them (as they feed and fill)
A Shepherd piping on a hill.   (40-45)

Not a great deal for us to *see* emerges from seventeenth-century poetry in any style. But Herrick's southern and western country, which differs so much from Cotton's northern, can be felt in its ages-old pastoral truth.

Even as Herrick was writing the poems gathered into *Hesperides*, newer and darker clouds began to appear. Herrick had long recognized the possibility of trouble, sometimes recommending virtue and sometimes enjoyment as a remedy. Enjoyment is prescribed in "A Paranaeticall, or Advisive Verse, to his friend, Master John Wicks."

Let's feast, and frolick, sing, and play,
And thus lesse last, then live our Day.
Whose life with care is overcast,
That man's not said to live, but last.   (30-33)

The anticipation of the cold season is yet stronger in another, even more Horatian poem, to the same close friend, "His age, dedicated to his peculiar friend, Master John Wickes, under the name of Posthumus."

And with a teare compare these last
Lame, and bad times, with those are past,
        While *Baucis* by,
My old leane wife, shall kisse it dry;
        And so we'l sit
By th'fire, foretelling snow and slit.   (81-86)

When the snow and sleet came, they brought winter to a way

of life. And in such hard times, Cavaliers often grasped at straws, mistaking a bright interval for summer. A Royalist like Herrick hoped so much for better times that he could think a Parliamentary reverse meant the return of good times.

> What gentle Winds perspire? As if here
> Never had been the *Northern Plunderer*
> To strip the Trees, and Fields, to their distresse,
> Leaving them to a pittied nakedness.

But the whole point of observing the signs of spring comes in the contrast between them and the certain evidence of political winter.[23] When one feels the full brunt of winter, one needs friends.

iii. The Winter Rites of Friendship

On 7 November 1644, that staunch Royalist James Howell was immured by the King's enemies in the Fleet. His letter of that date contains in little the significance of friendship in the Cavalier winter.

> Tho' the Time abound with Schisms more than ever, (the more is our Misery) yet, I hope, you will not suffer any to creep into our Friendship . . . You know there is a peculiar Religion attends friendship . . . There belonge to this Religion of Friendship certain due Rites, and decent Ceremonies, as Visits, Messages, and Missives.

[23] Herrick, "Farwell Frost, or welcome the Spring," ll. 9-12. Herrick's title is proverbial, a dismissal of that which one is glad to part with: see Morris Palmer Tilley, *The Proverbs in England* (Ann Arbor, 1950), F 769. The north wind probably refers to the Scots. The contrast of the seasons for psychological ends (in her face June, in her heart January) had long been a trope of love poetry. The perhaps related tropes of a false spring for a seeming change in political events, and of the winter of persecution, will be found in great detail in Dryden's fable of the Swallows in *The Hind and the Panther*, III. 415-638.

And then, referring to Castor and Pollux (though without the superb relevance in detail of Jonson's *"Dioscuri"* in his pindaric ode), Howell reminds his friend:

> You know that Pair which were taken up into Heaven, and placed among the brightest Stars for their rare Constancy and Fidelity one to the other: You know also, they are put among the *fixed* Stars, not the *erratics*, to shew there must be no Inconstancy in Love.

Then come the recurrent images of Cavalier poetry during the Interregnum.

> Navigators steer their Course by them, and they are the best Friends in working Seas, dark Nights, and Distresses of Weather.[24]

Other so-to-speak religions of these years had their rites and ceremonies, of course. In a fine love poem, "Day-Break," worthy of comparison with Donne's morning poems, Cotton makes the usual comparison of his mistress's eyes to the sun.

> Why should we rise t'adore the rising *Sun*,
> And leave the Rites to greater *Lights* undone?[25]

Love is the only religion I know of that has two deities, a goddess born from the sea and a god her son, who is blind. The religion of friendship, on the other hand, is one without

---

[24] Howell, *Epistolae Ho-Elianae*, II. xlvi (p. 352).

[25] Cotton, *Poems On Several Occasions* (London, 1689), p. 339; *Poems of Charles Cotton*, ed. John Beresford (London, 1923), p. 161. We must be grateful to Beresford and especially to John Buxton (*Poems of Charles Cotton* [London, 1958]) for their generous selections from Cotton's poems. But there are times when I at least should have preferred other choices and indeed the whole canon, including manuscript poems. Anyone interested in poetry owes himself the pleasure of reading through the almost 730 pages of the 1689 *Poems*, on p. 613 of which he will find, for example, "De Vita beata. Paraphras'd from the Latin."

gods. Friendship constitutes no idolatry, and to the minds of the Cavaliers it seems to have been the second religion of every country. If it seemed to lack prayer among its rituals, praise was certainly one of its "decent Ceremonies" and wine-drinking one of its "due Rites."

Since wine-drinking provides much for the happy life, and friendship another goodly portion, it should cause no surprise that the two combine with some political touches in Stanley's Anacreontic verse.[26] Other rites will be found in Stanley's *Register of Friends*. He tells how he and his uncle, and fellow Royalist poet, William Hammond, undertook their northern travels years before.

> But oh! the halcion-dayes we there convers'd
> Presag'd a storm, by which too soon dispers'd
> With equall grief, then, under different climes
> We wail'd the fury of unruly Times.   (13-16)

Another friend is Robert Bowman, whose Royalism lost him his position at Oxford and whose friend Stanley became when they met in exile abroad. In his apostrophe on this friend, Stanley once again speaks of the wars as a "Storm" (l. 5). In the apostrophe to Lovelace, the image is that of an "Eclipse" (l. 5). The rites with Bowman are those of conversation, and those for the dead Lovelace entail crowning with "both Laurells" of Pallas, "for Arts and Armes" (ll. 23-24). Stanley's *Register of Friends* has many attractions, but a sense of urgency is not among them. Writing as he does well after the Restoration,[27] he appropriately adopts a reflective tone. He had weathered the storm.

Before setting out for Paris and his meandering political

[26] See *The Poems and Translations of Thomas Stanley*, ed. Galbraith Miller Crump (Oxford, 1962), p. 92 (no. xxxix).

[27] Between 1675 and 1678, as James M. Osborn demonstrates in the important article mentioned in n. 5, above (see in it p. 4).

course, Waller celebrated Lucius Cary in a poem, "To my Lord of Falkland." Falkland had taken part in the fumbling expedition against the Scots in 1639. Like Waller on the Royalist side and Fairfax on the Parliamentary—like indeed Marvell on most sides—Falkland "misdoubted" (Marvell's word) his own cause. Unlike the rest, he could act even when sick from doubting and this capacity, with his great warmth, intelligence, and generosity must have made the visits by Waller and others to Great Tew one of the great satisfactions of the century. If formerly Jonson had been, in the town, the sun about which so many bright planets had danced, so Falkland gathered another constellation in the country. From that vantage point in one of the sweetest parts of Oxfordshire, some of the clearest understanding of the times emerged. Waller himself, like many other writers of the time, pictured the Grand Rebellion as the gigantom-achy (ll. 6-10), but he also regarded it as a storm.

> Some happy wind over the Ocean blow
> This tempest yet, which frights our Island so!
>
> (21-22)

The ceremony he longs to perform is giving Falkland a laurel crown.

It seems to have been Lovelace, crowned with the poet's and the soldier's laurels who, appropriately enough, most essentially dealt with the problems and the rites of the times. A poem examined early in this study, *Advice to my best Brother. Coll: Francis Lovelace*, counsels him on braving the "foul Winter" with strength of character. His "Sonnet. To Generall Goring, after the pacification at Byrwicke" cele-brates prematurely, and indeed wrongly, victory in the war. The night is past. Cups are to be filled and toasts drunk. It will be recalled that Lovelace's splendid Horatian ode (which I introduce in true but new guise for variety's sake), "The

Grasse-hopper. To my Noble Friend, Mr. Charles Cotton,"
comes to much the same call to wine, but with a very different
significance. What that significance is beyond what has been
said of the poem in previous chapters (and I may as well confess what must have been obvious, that I recur to this poem as
one example of the many-faceted character of Cavalier poetry), that significance can be understood in numerous ways.
First of all, it is a poem addressed to a friend whose name and
whose status as a friend are set forth in the title and repeated
in the poem. As the poem to Goring had looked outward upon
peace from the rites ending the war, so Lovelace here looks
inward after the vision of winter. An inward looking political
vision is perhaps the hardest to sustain, and the significance
achieved by such regard of experience certainly may verge
on the oblique. If we anticipate historically Marvell's *Horatian Ode* and set it beside this Horatian ode (for that is what
it is, though not so named by Lovelace), we shall see that
it was not Marvell who first sought direction by indirection.
Not only does Lovelace create the ideal polity in his own
breast and that of his friends, but also he does so with certain
crucial emblems (winter), rites (drinking), and formulas
(exchange of hearts) that the Cavaliers fitted together to
describe their case in the Interregnum. Among such motifs,
though not the first to command our attention, since its implications emerge only late in the poem, is that the grasshopper represents the royal. In *Anacreontea* (xxxiv), he is
βασιλεύς. In Stanley's translation (no. 43), the insect is
"Queen-like," and in Cowley a "*king*." Since Lovelace's poem
was long the most famous of the English Anacreontic poems,
we need feel no compunction to believe that Lovelace was
ignorant of what others knew.

Both Lovelace and Cowley go farther than Stanley or the
Anacreontic poet; that is, they see the grasshopper through
the summer of content and on in to wintry extinction. Cow-

ley does so by periphrasis: "Thou retir'st to endless *Rest*" (l. 34). Lovelace does so with far grimmer tone, and more complex art. In retrospect, we observe how the grasshopper's fall in the structural movement of the first half of the poem is enacted within the general downward movement of imagery in each stanza.

I

Oh thou that swing'st upon the waving haire
    Of some well-filled Oaten Beard,
Drunk ev'ry night with a Delicious teare
    Dropt thee from Heav'n, where now th'art reard.

II

The Joyes of Earth and Ayre are thine intire,
    That with thy feet and wings dost hop and flye;
And when thy Poppy workes thou dost retire
    To thy Carv'd Acron-bed to lye.

III

Up with the Day, the Sun thou welcomst then,
    Sportst in the guilt-plats of his Beames,
And all these merry dayes mak'st merry men,
    Thy selfe, and Melancholy streames.

IV

But ah the Sickle! Golden Eares are Cropt;
    Ceres and *Bacchus* bid good night;
Sharpe frosty fingers all your Flowr's have topt,
    And what sithes spar'd, Winds shave off quite.

The first three stanzas carry most of the Anacreontic burden, except of course that they lead to the mortal "night" of the fourth. Such a rhythm (which will be played in a contrapuntal fashion in the second half of the poem) is gently enacted by the conclusion of each of the first three stanzas, although the "night" of these three becomes progressively less dark until the fourth stanza. We see such subtle movements in the reference in the first stanza to the Grasshopper's drunk with dew "Dropt thee from Heav'n," in the insect's turning to bed in the second stanza; and, in the gentlest of touches, in the soft playing of "Melancholy streames" in the third. The climax of the poem comes of course with the fifth stanza.

> Poore verdant foole! and now green Ice! thy Joys
> Large and as lasting, as the Peirch of Grasse,
> Bid us lay in 'gainst Winter, Raine, and poize
> Their flouds, with an o'reflowing glasse. (17-20)

Night, winter, storms, and floods—these emblems of political upheaval surely appear to enter with a naturalness that one associates with Cotton's "catholic decorum," with his ability to use natural detail as if for itself and without its usual emblematic freight. Here we face the central problem of the poem.[28] It will be clear that I follow the political interpretation of the poem, but the weighing and adjustment of such an interpretation seems crucial. The social implications cry out to me in "The Ant," "The Snayl," and "The Toad and Spyder" and they elude me in *The Falcon*. "The Grasse-hopper" speaks out, but with some elusiveness.

Too much in the poem speaks to ends other than those of the insect world. Or, more accurately, the insect holds emblematic properties. To argue by analogy from lore surrounding another, more often treated insect, bees possessed emblematic significance in terms of kingly government and civil

---

[28] The lesser problems are not few. Why does Lovelace change line length in the second lines of the second, fifth, and sixth stanzas? And what insect is involved? Should we think of the cicada of Liddell and Scott or the cricket of the Loeb edition (*Elegy and Iambus*, vol. 2, ed. J. M. Edmonds)? The classical learning deployed on these sixty-odd poems since Stephanus's *editio princeps* in 1554 quite catches the breath.

war. And such emblematic strains by no means fade out during the seventeenth century. Thomas Moffett's (or Muffet) *Theatre of Insects* was added to Edward Topsell's *History of Four-footed Beasts* in its 1658 edition. Dryden shows himself acquainted with such lore, and in the eighteenth century there were two enormously popular books: *The True Amazons: or, the Monarchy of Bees* by Joseph Warder and ΜΕΛΙΣΣΗΛΟΓΙΑ, *or, The Female Monarchy* by John Thorley. In short, I take it that Lovelace's insect represents indeed the Grasshopper King, but that there is no need fully to identify the two or to assume that all things true of one must be true of the other. But nothing in either the literary traditions or the history of the century leads me to shrink from thinking Charles I a "Poore verdant foole." Only, he was not then "green Ice." In the first half of the poem that dismissive gesture is balanced by apotheosis, by the Grasshopper King's being "reard" to heaven (1. 4). In this instance, apotheosis proves more appropriate for the King than the Grasshopper. In addition, we do not lack other images with Royalist potential: the sun (1. 9) and gold (1. 10). And further, one of the subtlest movements in the first half of the poem comprises those rites of the day and of the year that Charles Cotton the Younger was to celebrate in the next generation. With Lovelace, those rites are also the rites of a lifetime: for the Grasshopper and, with due subtractions, for Charles I.

The second half of the poem plays off against such largely negative views of the state (and of experience in it) a positive view of another polity, friendship. In the crucial fifth stanza, the Grasshopper's fate should lead ("Bid") the friends to poise against the floods of winter and rain their own flooded glasses of wine. In the next stanza, contrary to "this cold Time and frosen Fate," the friends will create a "Genuine Summer in each others breast." The crucial word for following Lovelace's dim tracks seems to me to be "Genuine." One could perhaps toy with the Latinate meaning of *ingenuus* ("freeborn"), but in fact the word centrally means "authentic," "real," as opposed to the false or at least transitory summer of the Grasshopper King. (The problem is that the insect's summer *was* "Genuine" by the calendar.) As the rest of the poem shows, the scene remains winter, so that "Genuine" can only mean that the metaphorical meaning (the "tenor," the friends' so-to-speak summer as opposed to the "real" summer just passed) is taken to be more "Genuine" than the authentic summer lately fled. And so we are led on to discover that the polity of friendship excels that of monarchy itself.

> Thus richer then untempted Kings are we,
>   That asking nothing, nothing need:
> Though Lord of all what Seas imbrace; yet he
>   That wants himselfe, is poore indeed.   (37-40)

No names are named; but more importantly, it has been five stanzas since the Grasshopper was mentioned. The Grasshopper King was glorious. Yet he had no friends. That lack receives no stress and the deficiency is nowhere dwelt upon. But what a contrast we sense between the solitary, happy insect perishing with the cold and the friends suffering from the rigors of political winter but happy in their own summery polity of friendship. If one swallow does not make a spring, one grasshopper will keep no summer. But two friends create their own summer in the dead of wet, blustery, dark winter.

Does an air of escapism hang over such assertions of triumph? Did Lovelace really think that he and Cotton could create a *genuine* summer in the midst of a metaphorical winter? Such questions must finally be left to the reader or in limbo, Lovelace being dead this long while. But to follow his indirections, my answer would be, not yes or no, but Friendship. We are back with Aristotle on the necessity of concord in the *polis,* and friendship as the best example of civic concord. From the beginning of Lovelace's poem, we observe its macrocosmic character in spite of its celebration of the Anacreontic insect. We are given heaven and earth in the first two stanzas, and then for three that cosmic force, time. Against that force, as we have often seen, the Cavaliers finally had to pitch their greatest hero, the good man: "Thou best of *Men* and *Friends.*" Whether we wish to retain the superlative for the *vir optimus* is not so important as that we observe that, given such united enemies as time and the times and they attacking on so wide a scale, Lovelace counters not only with a good man but a good man who is a good *friend.* The concord that they will bring, simply by being friends and observing the drinking rites of friendship, reappears in an emblematic conceit: the blazing hearth singes the frosty wings of the North Wind, and the scene is "*Aetna* in Epitome." Or, in Cowley's conceit for concord in his pindaric ode to Hobbes, "So *Contraries* on *Aetna's* top conspire" (st. vi).[29]

In the end, however, it is not so much emblems or certain rather Herrick-like touches ("sacred harthes," "Vestall Flames," "cleare *Hesper*") that enable the poem to succeed in its aim. What proves essential can be pointed to most easily in that part of the poem that always brings me a degree of shock. Throughout the first half of the poem, the "Thou" addressed is the Grasshopper. Then suddenly, it is Cotton, "Thou best of *Men* and *Friends.*" It always almost seems as if Cotton succeeds to the Grasshopper's role in the scheme of things as well as in the verse apostrophe. The shift accompanies the claim that the friends can make a genuine summer, but it also forecasts a double shift in one aspect of the Grasshopper. (Actually, to be fully accurate but perhaps

[29] Cowley's note to the sixth stanza gives some of the classical references.

obscure, this shift leads to two further stages of the poem in which differing forms of kingship are conveyed, forms that finally and *retrospectively* define the kind of monarch the Grasshopper King has been.) The friends at their hearth in that genuine summer observe that "Dropping *December* shall come weeping in" (l. 29). His flood of tears will be poised (to adopt an earlier usage altogether similar) by the "show'rs of old Greeke" that the friends drink. Although his reign has been usurped, those friends who have created a genuine summer will also show themselves able to give December "his Crowne againe!" (l. 32). Once more we meet that offensive Parliamentary ordinance against celebrating such feasts as Christmas. At the "sacred harthes" of Cavalier friends, the old true religion (as well as the old true pagan celebrations that accompanied it) is restored. Outside, in the winter of the state, on the other hand, King Christmas has been usurped. I say "King Christmas" in a sense analogous to "Father Christmas," the English Santa Claus, partly because that seems to be the median between the metaphorical vehicle (December as king) and the metaphorical tenor (Christ the King). But it also brings us the second king of the poem and leads us to see that in the end the friends are not only "richer then untempted Kings" (l. 37) but also are regal themselves. What Cotton shares with the Grasshopper, then, is what he shares with December and with Lovelace, kingship. The role of each in the sharing finds definition through friendship.

If such a discussion proves acceptable, certain conclusions follow. In senses limited at every point, the poem presents us with three realms—those of the Grasshopper King, of King Christmas and of the Royal Friends. The first is deposed or executed by improvidence, nature, and time. The second is restored by the friends. And the friends prove themselves true kings by ruling themselves, which is, as we have seen, an old commonplace.[30] The values of monarchy are restored by a friendship that creates on a small scale the self-sufficient, moral good life. (With a few "old Greeke" touches of the happy good life.) It turns out that dominion over John of Gaunt's sceptered isle is not the true royalty; the true is self-possession:

> Though Lord of all what Seas imbrace, yet he
> That wants himselfe, is poore indeed.   (39-40)

My interpretation of Lovelace's splendid poem suggests that his Grasshopper provides us with a limited but suggestive *Eikon Basilike*, an image of a king. There was of course published during those times the great Royalist propaganda triumph of the century, *Eikon Basilike* (1648). Attacked mordantly by Milton, it was defended ardently by Royalists. For our purposes, one of the most interesting by-products of that whole image-crisis will be Thomas Stanley's *Psalterium Carolinum*. Since the collection is dedicated "To His Sacred Majesty Charles the Second," it must have been prepared sometime between the execution of Charles I in 1649 and the date of publication, 1657. In the spirit of Lovelace, the ardent Royalist poet restores King Charles by translating for the second Charles the prayers attributed to the first in *Eikon Basilike*, with of course allusion to the usual Stuart type, David. My aim in what follows does not include any suggestion that Lovelace had *Eikon Basilike* in mind as he wrote his poem. Rather, I hope to suggest that there were widely common images, or emblems, and widely common ways of handling them to convey the vicissitudes of kings. I believe that a minimum of commentary will suffice. Ode IV is "Upon the insolency of the Tumults."

[30] On the maxim, "Sibi imperare imperium maximum," see ch. iii, n. 34.

The floods, the floods, o're-swell their bounds,
Danger my threatned soul surrounds.
Mine and my Realms iniquity,
(The tumults of our souls 'gainst thee)
These popular inundations cause,
That bear down Loyalty and Lawes.   (5-10)

Ode vii, "Upon the Queens departure and absence out of England," brings other familiar images with conquests of time and darkness.

May knowledge of Earths vain delights,
Ecclips'd by unexpected Nights,
    By sudden Stormes ore-cast;
Enflame our Spirits with desire,
To those Celestiall joyes t'aspire,
    Which time shall never wast.   (37-42)

Ode xix, "Upon the various events of War, Victories, and Defeats," shows, like Lovelace, how one flood may poise another, the blood shed in war.

To me impute not, Lord! the purple Flood,
    Shed with unwilling grief in my defence.
But wash me in my Saviours precious blood:
    By whom my troubles hope a quick dispence;
    For short are impious joys, and Confidence.
                                    (31-35)

And the observance of rites, which is of course implied by the whole *Psalterium*, receives specific comment in Ode xxiv, "Upon their denying his Majesty the attendance of his Chaplains."

And scatter'd like a dying Coale, from all
  Those pious glowings that might fire impart:
  Keep and increase on th'Altar of my Heart,
On Thee in sacrifice of Pray'r to call.   (25-28)

The analysis of the problems of Charles I and his cause in these "psalms" resembles that in Lovelace's poem. The imagistic terms and procedures agree at many points. Since numerous other poems might be adduced to show similar resemblances in treating the same general topic, we can see that the Cavaliers possessed a means to understand their plight. The story behind *Eikon Basilike* is a tragic one (let us grant that Charles proved great in adversity), as Stanley's odes show very well. In "The Grasse-hopper," however, Lovelace solves the same problems, and triumph replaces tragedy. How this can be is once more to be explained with one word: friendship. Like the Grasshopper King, the king of *Eikon Basilike*—and especially in the frontispiece and prayers—is a solitary figure. Lovelace and Cotton make up, in friendship, a real and just society that does not require laws: a perfect state does not need them. Instead it is upheld by the creative rites of friendship.

Among the numerous variations rung on such harmonies, we may appropriately briefly consider three poems by Charles Cotton the Younger, as a kind of repayment to Lovelace for praise of Cotton's father. In "The Storm. To the Earl of ——," the relation of a storm at sea is prefaced by the assurance that friendship will keep all things right. (The poem seems to treat an actual return crossing of the English Channel and concerns friendship without observable political emphasis.) In "The Tempest," the stilling of the storm is a matter raised, as it usually was raised, at the end of the poem. But here love rather than friendship promotes concord.

Finally, we may recall *Winter*, that personified enemy consigned off to the cold and political north.

> Or, let him *Scotland* take, and there
> Confine the plotting Presbyter;
> His Zeal may Freeze, whilst we kept warm
> With Love and Wine, can know no harm.
>
> (st. liii)

Since winter was more likely to seize Scotland before Derbyshire, the image seems illogical. That is, in the seventeenth-century winter is naturally depicted as coming from the north, not going to it. But of course the Cavalier winter was only partly real, and I believe that Cotton alludes to the situation in Scotland in 1650-1651. "Old Winter" would then be an emblem of the Commonwealth, and particularly its armies under Cromwell, who in fact did go in the years mentioned into Scotland to "Confine the plotting Presbyter." The immediately preceding stanzas show how the incursion by winter may be defeated by celebrating the rites of friendship in drinking and by making love. The same self-sufficiency that characterizes Lovelace's conclusion comes also in Cotton's poem: "What would we be, but what we are?" (st. xlix). The rites of the friends will lend succor even to those driven abroad by the hard times: "Men that remote in Sorrows live, / Shall by our lusty Brimmers thrive" (st. lxv). So much for winter.

Much of the furor over the wars and revolutions in England in the seventeenth century appears at this distance to have resulted from battles of words. It does seem arguable, that by fighting with paper, the contentious parties saved themselves blood. And so a metaphor like winter could both define the problem of the Cavaliers and assist them in solving it.

It seems appropriate to conclude this stage of our concern

with friendship with the poem Cotton wrote "To the Memory of my worthy Friend Colonel Richard Lovelace," which was appended to *Lucasta. Posthume Poems* in 1660. Lovelace had found it necessary to take refuge about 1655 in what the younger Cotton termed "our little Castle"—fortified by wine. And we may presume that they observed the rites that Cotton's father had practiced earlier with the colonel. Cotton manages very handsomely to praise friendship and friends, but he attributes the conquest of time to "*Vertues*" and "Merits" possessed by Lovelace.

> And though thy *Vertues* many Friends have bred
> To love thee Living, and lament thee Dead,
> In *Characters* far better coucht than these,
> Mine will not blot thy *Fame*; nor theirs increase;
> 'Twas by thine own great Merits rais'd so high,
> That, maugre *Time*, and *Fate*, it shall not die.
>
> (41-46)

In the year of the Restoration, Ben Jonson's music of virtue was once more played. And it is now to those years about the end of the Protectorate and the beginning of the Restoration that we turn.

iv. Country Hopes of Spring

The bravado of poems like *Winter* and "The Grassehopper" seems to exaggerate the relief that drinking in any literal sense might afford to the Royalists. But if we take those lifted "Brimmers" and blazing "harthes" as they were meant, we will see that they make sense. Those Cavaliers who were not altogether dispossessed of their lands could indeed retreat to country estates. There, with their friends, they could cheer each other through the cold season, maintain their morale, *write* at least against the government, and hope for a reviving spring. The concord of friendship (let it be said for the last time) was considered the basis of the state by writers from Aristotle on. Much else was implied by way of matters social, religious, and artistic by those Cavalier friends in the country. For in an important sense, the rites of Cavalier friendship preserved the Constitution until it was once again the friends' turn to occupy the posts of church and state.

The bravado I have spoken of seems to yield, as time goes on, to patience or at least resignation, and then discernibly to hope that the winter was over and gone. It should be confessed at once that I have no more evidence of dates of composition in mid-century than the next man, and I may be describing varying rather than successive moods. But there does seem a difference worth observing between Cotton's *Winter* and one of his nicest, and also politically resigned, poems of compliment. "To my dear and most worthy Friend, Mr. Isaac Walton," which begins in the usual lament about the season.

> Whilst in this cold and blust'ring Clime,
>   Where bleak winds howl, and Tempests roar,
> We pass away the roughest time
> Has been of many years before.

The very cyclical pattern of the seasons—those rites of time that we observe in his diurnal quatrains—brings that hope all men can share when things are at their worst. That is, he finds that there is

> some delight to me the while,
>   Though nature now does weep in Rain,
> To think that I have seen her smile
> And haply may I do again.   (15-18)

Cotton invites Walton to fish in his beloved Dove on "A day without too bright a Beam, / A warm, but not a scorching

Sun" (ll. 29-30), and so the friends will think themselves happier than the great Leviathans of the world. Whether Cotton alludes to Hobbes we cannot absolutely say, but we can be sure that Cotton has found out a way to live with winter and even to avoid the extremes of a summer regime. The way is the way of friendship, and the prospect of being with Walton and Cotton along the Dove would put anyone into boots.

The angling rites of friendship are less common than the drinking ones. But conversation provided other rites, as Howell said, and as we can see from a poem perhaps written two or three years before Cotton's piscatory epistle to Walton. That is, if Cotton can be imagined to have written that poem sometime soon after *Leviathan* came out in 1651, Vaughan was probably writing in the late winter of 1649, after the execution of Charles I in January, *To his retired friend, an Invitation to Brecknock*.[31] The first forty or fifty lines of direct address have some felicities but little intensity. (We observe in Vaughan's Cavalier poetry the same fluctuating poetic pressures that we discover in his Metaphysical poems.) But, if one may say as much of a poem from a collection entitled *Olor Iscanus*, lines 55 to 88 strike true fire.

> Come! leave this sullen state, and let not Wine
> And precious Witt lye dead for want of thine.
>
> (55-56)

And, in one of the most memorable natural images of a whole century of poetry, Vaughan introduces the Cavalier winter.

> Come then! and while the slow Isicle hangs
> At the stiffe thatch, and Winters frosty pangs

[31] See French Fogle, ed., *The Complete Poetry of Henry Vaughan* (New York, 1965), p. 64, n. 9.

Benumme the year, blith (as of old) let us
'Midst noise and War, of Peace, and mirth discusse.

(73-76)

We observe here the specification of the meaning of the
Cavalier winter, as well as the slow icicle. Vaughan sounds a
true mid-century note (heard so often in Cotton and Cowley
as well) of "contentation" (Cotton's word) to stay retired
in his country village: "Why should wee / Vex at the times
ridiculous miserie?" (ll. 77-78). There seems to be real confidence in the conclusion of the poem.

Innocent spenders wee! a better use
Shall wear out our short Lease, and leave th'obtuse
Rout to their *husks*; They and their bags at best
Have cares in *earnest*, wee care for a *Jest*.   (85-88)

Vaughan's content in a circle of country friends did not
prevent his reaching other circles beyond his village. The
most remarkable of those with which he came into contact
in the years before and just after the Restoration was no
doubt that of Katherine Philips, "the Matchless Orinda."
The cult, ceremony, and vicissitudes of friendship were developed by her circle with a "Platonic" purity and a degree
of rarefied success unattained by Henrietta Maria. Throughout this chapter I have been stressing the importance of
circles of friends to the milieu and the self-understanding of
the century. And Mrs. Philips's group reminds us of the Tribe
of Ben, of Little Gidding, of Great Tew (or of Bunyan's
Baptists at Bedford or Dryden's wits at Will's). Perhaps the
comparisons sound forced? Certainly I should not force her
verse on those who do not share my passion for the seventeenth century at all stages, but here is at least the title of
one of her poems: "To the Excellent Mrs. Anne Owen, upon
her receiving the name of Lucasia, and Adoption into our

Society, Decemb. 28. 1651." The ritual recalls Jonson sealing one of the Tribe of Ben, and it foretells Dryden offering snuff to favored young writers. Whether Orinda might not have expired instantly in the presence of those two poets may well be wondered, but her ceremonies of friendship resemble theirs in the most essential particular: in being a ceremony of friendship.

Orinda expresses no fervent Royalism in her poems, but she knew the troubles, the hopes, and the themes we have been following. "A Country-life" is yet another of the variations on Horace's second epode. As such, and in spite of its declaration that "Friendship and Honesty" are the only truly good things on earth,[32] the technique of this country poem is rather that of statement than the address to a friend usual in her verse. Orinda also differs from the main line of the Cavaliers and their ladies in her austerity. She definitely would make Vaughan a better friend than she would Cotton.

> Let some in Courtship take delight
> And to th' *Exchange* resort;
> There Revel out a Winter's night,
> Not making Love, but Sport.
>
> (p. 180)

We must understand that "making Love" concerns the soul alone; "Peace and Honour" (p. 180) are her claim.

> In this retir'd Integrity,
> Free from both War and noise,
> I live not by Necessity,
> But wholly by my Choice.
>
> (pp. 181-82)

This ending closely resembles that of Vaughan's "To his retired friend," and since we can move from that to Cotton in-

---

[32] Katherine Philips, *Poems* (London, 1664), p. 181.

viting Walton and to the friendly gestures by numerous other poets, we can see that some at least of the aerial, which is to say rather thin-blooded, poetry of the Matchless Orinda practices the Cavalier mode.

The resemblance between his and Mrs. Philips's poetic close may return us to Vaughan's response to her death in 1664, after Charles II had returned from his travels. Vaughan thinks that those poets who had thrived in the hard times now succeed less well.

> For since the thunder left our air
> Their *Laurels* look not half so fair.[33]

Nonetheless, the publication of Orinda's poems tells us that at least the hard winter has passed.

> So while the world his absence mourns
> The glorious Sun at last returns,
> And with his kind and vital looks
> Warms the cold Earth and frozen brooks:
> Puts drowsie nature into play
> And rids impediments away,
> Till Flow'rs and Fruits and spices through
> Her pregnant lap get up and grow.[34]

The royal sun had indeed returned; and with a prescience perhaps owed to his brother's Hermeticism, Vaughan here

[33] "To the Editor of the matchless Orinda," ll. 17-18. As French Fogle notes, the "thunder" is that of the Civil Wars. It was also believed that laurel protected the head from thunderbolts, an idea that Marvell plays with in *An Horatian Ode*, ll. 23-24.

[34] Lines 25-32. One sees little apparent reason for extending the simile *except* to comment on the change of the times. A better elegy still is that *To the pious memorie of C. W. Esquire*. I think Vaughan an uneven poet but a fine one, and I hope it is plain that I share with E. L. Marilla admiration of Vaughan's secular as well as religious poetry: see "The Secular and Religious Poetry of Henry Vaughan," *Modern Language Quarterly*, IX (1948), 394-411.

predicts Charles's remarkable fertility, in so many areas of life and thought. As another poet expressed it,

> The Royal Husbandman appear'd,
> And Plough'd, and Sow'd, and Till'd,
> The Thorns he rooted out, the Rubbish clear'd,
> And blest th' obedient Field.[35]

By two of those coincidences that meant so much in the seventeenth century, Cromwell died as high winds blew, whereas the month of Charles's return to end the Cavalier winter was May. On the 25th the royal husbandman landed at Dover. He entered London on the 29th, which was his thirtieth birthday.

> The rising Sun complies with our weak sight,
> First gilds the Clouds, then shews his globe of light
> At such a distance from our eyes, as though
> He knew what harm his hasty Beams would do.
>
> But your full *MAJESTY* at once breaks forth
> In the Meridian of your Reign . . .

So Waller, blinking his eyes, hardly believing that the emblematic royal sun was again warming the world.[36] But poets and critics have been known to exaggerate, even if Waller confessed to Charles that this poem was inferior to his other on Cromwell because poets succeed better in feigning than in telling the truth. All the poets mentioned in this section survived into the Restoration (even Horace, so to speak), and just as much of the finest Cavalier verse was written during the first two decades of the Restoration, so was most of

[35] Dryden, *Threnodia Augustalis*, ll. 356-59. Cf. also "To my Dear Friend Mr. Congreve," ll. 6-10 and, of course, the opening lines of *Absalom and Achitophel.*
[36] Waller, "To the King, upon his Majesties Happy Return," ll. 1-6.

the Cavalier poetry of mid-century republished or published for the first time after 1660.

That being so, I shall leave Orinda (whom I do not much boast of) to end my chapter on that most attractive and steadfast of Cavalier enterprises, friendship, with some lines from a poem published in the fourth edition of Izaak Walton's *Lives*. Dated "Jan. 7, 1672," Cotton's stanzas "To my Old and most Worthy Friend, Mr. Izaak Walton, on his Life of Dr. Donne, etc." wonderfully capture something of the whole century and embody the quiet appeal to us of the relations between these two superlative anglers. Walton had been a friend of Cotton's father and had preserved the memories of two of his father's other friends.

> How happy was my Father then! to see
> Those men he lov'd, by him he lov'd, to be
> Rescu'd from frailties, and mortality.
>
> *Wotton* and *Donne*, to whom his soul was knit:
> Those twins of Vertue, Eloquence, and Wit,
> He saw in Fame's eternal Annals writ.   (25-30)

Walton is the truest friend.

> For whereas most mens friendships here beneath,
> Do perish with their friends expiring breath,
> Yours proves a Friendship living after Death;
>
> By which the generous *Wotton*, reverend *Donne*,
> Soft *Herbert*, and the Churches Champion,
> *Hooker*, are rescued from oblivion.   (49-54)

We even discover woven in this lovely design many of the strands of imagery that have run through this chapter, as Cotton speaks of Wotton's service to the state and Donne's brilliant achievements. Walton was more than a fair-weather friend:

Nay, through disgrace, which oft the worthiest have;
Through all state-tempests, through each wind and wave,
And laid him in an honourable grave.

And yours, the whole worlds beloved *Donne*,
When he a long and wild carere had run
To the meridian of his glorious Sun . . .   (76-81)

The chains of friendship and of friends in the seventeenth
century are as long and as complex and as crucial in determin-
ing the genesis of Cavalier poetry as is nucleic acid for the
coding of the human body. Walton treated Cotton as his
son and his friend. And at the beginning of the second part
of *The Compleat Angler*, when Piscator (Cotton) encounters
Viator, he strikes up a friendship at once on learning of their
common love for Master Walton. Piscator speaks with great
pride of their friend: "I must tell you, further, that I have
the happiness to know his person, and to be intimately acquainted with him, and in him to know the worthiest Man,
and to enjoy the best, and the truest Friend any Man ever
had."[37] No small portion of Cavalier poetry, and of other
writing of the seventeenth century, is devoted, if seldom with
such unaffected candor as this, to the Compleat Friend.

[37] Cotton, *The Compleat Angler* (Part II) (London, 1676), ch. i.

POSTSCRIPT

THE IMAGE OF Izaak Walton and the younger Charles Cotton
fishing in the Dove on a fine day conveys something of those
golden moments that have made Cavalier poetry appealing to
so many and for so long. And the sentence just written offers
a précis of parts of this book in the order those parts appear.
Rather than parse one of my own sentences, however, I would
like to reexamine a characteristic passage by Ben Jonson. He
is addressing a noblewoman.

> Your happy fate,
> For such a lot! that mixt you with a state
> Of so great title, birth, but vertue most,
> Without which, all the rest were sounds, or lost.
> 'Tis onely that can time, and chance defeat:
> For he, that once is good, is ever great.

We *know* that this can only be written by Jonson or someone
unusually able to rise to his level and following his example.
We *know* that this is not Donne in one of his poems to Lucy,
Countess of Bedford, or Dryden addressing the Duchess of
Ormond. To explain why we know this, I sought in the first
chapter to establish if possible the reason in terms of the
mode of poetic address. The phrase used was "the social
mode." Jonson stands, as it were, in society but not in public,
so distinguishing him from Dryden. Jonson speaks in a tone
appropriately overheard by others, rather than with a tone of
intense devotion to a single person, so distinguishing him
from Donne. All three poets would agree that social title,
good birth, and virtue are standard topics of praise, and that
the superiority of virtue to other endowments is also a commonplace. In their ways, Donne and Dryden touched the
same commonplaces in writing praise of women. Although
much else besides their angle of vision differed, surely it is the

radical mode of presentation, the aesthetic distance, or simply the stance that differs among the three poets. There have been times when each of these writers has suffered for lack of appreciation, indeed for lack of understanding, of their ways of regarding themselves in relation to other people. And therefore, although numerous alternatives no doubt exist to my phrase, the concept underlining "the social mode" seemed to me to require introduction at the very outset of this book.

The lines from Jonson's poem obviously celebrate an ideal. Poets who aim at such celebration are faced with the problem of bringing nearer, or of making actual, the remote possibility of what is ideal. The phrase, "so great title, birth, but vertue most," depends upon a conception of society, as I have already mentioned, and on a conception of ethical norms. The advantages of social distinction on the one hand and of personal goodness on the other do indeed produce a "happy fate." Such conceptions clearly relate to the central ideal in Cavalier poetry, the good life. In the second chapter and subsequently I have explained at length how the happy or blessed life and the morally good life differ and how they might be related. As we can see in Jonson's passage alone, the distinctions between the two kinds of good life must be made in order that they may have other distinctions added to modify or transcend them. His noblewoman is a "great" lady, in several senses of that term. But the adjective appears again in that splendid last line: "For he, that once is good, is ever great." The morally good life produces a greatness superior to the accidentals of birth, and it is a possibility open to each man. By the same token, that greatness superior to greatness of social class and rank is also granted to Lady Aubigny, making her a model for her virtue as well as a lady "happy" in her "state."

Jonson's passage also takes us on our next step in telling us of the threat to the good life and of the means to overcome the threat. If only "vertue" can defeat time, then time is the worst enemy. Put another way, "he, that once is good, is ever great." What I termed "The Ruins and Remedies of Time" for a lengthy chapter, Jonson compresses into essential poetry. All this needs no special detail at this stage, but if I may stress from those lines "happy fate," "time," and "chance," we can see that the order of fate and the order of time differ here, one order being good and one bad. But both differ from chance, which is haphazard, unordered. For this reason, and because of the assumptions involved in such a ubiquitous seventeenth-century phrase as "art and nature," I felt it necessary to move from the chapter on time and the times as threats to the good life to "Order and Disorder." It was necessary, I felt, to distinguish the ways in which one or the other could be a value or a threat, could be distasteful or welcome. In one sense, what I had to say about the good life was complete only when I had written about the perceptions of time, order, and disorder that the Cavaliers realistically and maturely knew to bear directly on their central ideal.

I had proceeded so far fully realizing that some readers would have been wondering when I would get around to talking of two major poetic subjects connected with the good life in Cavalier poetry: love and friendship. There is an enormous literature, both imaginative and discursive, on both these subjects in literature and very often, from Plato forward, they have been treated together. It seemed wisest, therefore to examine love, after "Order and Disorder," since as we have seen in the fifth chapter, the Cavaliers and their contemporaries had considerable difficulty in sorting out the ways in which love was indeed a species of order or of disorder. By separating from love the other subject of value, friendship, and by making it the subject of the last chapter, I was able to close with an unquestioned positive norm in the good life. Jonson recognized that there might be serious doubts as to whether he was a love poet in the usual sense, whether his poems carried conviction that he was himself in love. But he leaves us with no doubt whatever that he is our major poet of friendship and that his contemporaries saw the value as much in the man as in the poet. Because experience changed between Jonson and Cotton, and because talents differ in kind as well as degree, poetry by different hands alters in the years after Jonson's triumph in creating a new poetry. And although the historical pattern of development has not been a major concern in this book, the reader will have put all in chronological place even before I have mentioned the desirability of that being done.

I hope such a reader will forgive my adding one more thing, although it has already been implied. That conception of the good life that is central to Cavalier poetry is in one sense wholly traditional and it certainly does not end with them. It is at the heart of much classical philosophy, especially that of Aristotle and Cicero, but also in major respects of the ideas of Plato and Seneca as well. In another sense, it is a concept central to Christianity. To say that Jonson created a new poetry and thereby a new vision of life is therefore to say that from something old and from elements enduring in human experience he kindled individual life in that which was moribund or familiar. And as a literary artist he was able to envision many, although not by any means all, of the artistic means of creating a poetry consistent with his vision. That is an artistic virtue of which it may also be said that " 'Tis onely that can time, and chance defeat."

When therefore one says that Ben Jonson is a great poet, one means something about his art. One also means something about his values recreated by his art. Readers of *The Metaphysical Mode from Donne to Cowley* know the strength of my conviction that the same is true for other poets of that time. One is also not denied opportunities to show the same for those other great poets, principally Milton and Dryden, who brought yet another kind of art and another view of life to seventeenth-century poetry, and who bring those things to us in their writing. By the time my son and daughter grow old enough to read my seventeenth-century studies appreciatively, they will no doubt find my modes out-moded. But I wish them to be able to say, with whatever allowance they must make for me, that their father found some happiness in seventeenth-century poets and that the values of those poets contributed to his own sense of the good life.

APPENDIX
THREE POEMS DISCUSSED
AT LENGTH

i. Edmund Waller, "At Pens-hurst"

While in the Park I sing, the listning Deer
Attend my passion, and forget to fear.
When to the Beeches I report my flame,
They bow their heads as if they felt the same:
To Gods appealing, when I reach their bowrs   5
With loud complaints, they answer me in showrs.
To thee a wild and cruel soul is given,
More deaf than trees, and prouder than the heaven.
Loves foe profest, why dost thou falsely feign
Thy self a *Sidney*? from which noble strain   10
He sprung, that could so far exalt the name
Of Love, and warm our Nation with his flame,
That all we can of love or high desire,
Seems but the smoak of amorous *Sidneys* fire.
Nor call her mother, who so well do's prove,   15
One breast may hold both Chastity and Love.
Never can she, that so exceeds the spring
In joy and bounty, be suppos'd to bring
One so destructive; to no humane stock
We owe this fierce unkindness, but the rock,   20
That cloven rock produc'd thee, by whose side
Nature to recompence the fatal pride
Of such stern beauty, plac'd those healing springs
Which not more help than that destruction brings.
Thy heart no ruder than the rugged stone,   25
I might like *Orpheus* with my numerous moan

Melt to compassion; now my traitrous song,
With thee conspires to do the Singer wrong:
While thus I suffer not my self to lose
The memory of what augments my woes: 30
But with my own breath still foment the fire
Which flames as high as fancy can aspire.
    This last complaint th'indulgent ears did peirce
Of just *Apollo*, President of Verse,
Highly concerned, that the Muse should bring 35
Damage to one whom he had taught to sing:
Thus he advis'd me, on yon aged tree,
Hang up thy Lute, and hye thee to the Sea,
That there with wonders thy diverted mind
Some truce at least may with this passion find. 40
Ah cruel Nymph from whom her humble swain
Flies for relief unto the raging main;
And from the windes and tempests do's expect
A milder fate than from her cold neglect:
Yet there he'l pray that the unkind may prove 45
Blest in her choice, and vows this endless love
Springs from no hope of what she can confer,
But from those gifts which heaven has heap'd on her.

ii. Edmund Waller, *On St. James's Park as lately improved by his Majesty*

Of the first Paradise there's nothing found,
Plants set by heav'n are vanisht, and the ground;
Yet the description lasts; who knows the fate
Of lines that shall this Paradise relate?
    Instead of Rivers rowling by the side 5
Of *Edens* garden, here flowes in the tyde;
The Sea which always serv'd his Empire, now

Pays tribute to our Prince's pleasure too:
Of famous Cities we the founders know;
But Rivers old, as Seas, to which they go,   10
Are natures bounty; 'tis of more renown
To make a River than to build a Town.
For future shade young Trees upon the banks
Of the new stream appear in even ranks:
The voice of *Orpheus* or *Amphions* hand   15
In better order could not make them stand;
May they increase as fast, and spread their boughs,
As the high fame of their great Owner grows!
May he live long enough to see them all
Dark shadows cast, and as his Palace tall.   20
Me-thinks I see the love that shall be made,
The Lovers walking in that amorous shade,
The Gallants dancing by the Rivers side,
They bath in Summer, and in Winter slide.
Me-thinks I hear the Musick in the boats,   25
And the loud Echo which returns the notes,
Whilst over head a flock of new sprung fowl
Hangs in the ayr, and does the Sun controle:
Darkning the sky they hover or'e, and shrowd
The wanton Sailors with a feather'd cloud: 30
Beneath a shole of silver fishes glides,
And playes about the gilded Barges sides;
The Ladies angling in the Crystal lake,
Feast on the waters with the prey they take;
At once victorious with their lines and eyes   35
They make the fishes and the men their prize;
A thousand Cupids on the billows ride,
And Sea-nymphs enter with the swelling tyde,
From *Thetis* sent as spies to make report,
And tell the wonders of her Soveraign's Court.   40
All that can living feed the greedy Eye,

Or dead the Palat, here you may descry,
The choicest things that furnisht *Noahs* Ark,
Or *Peters* sheet, inhabiting this Park:
All with a border of rich fruit-trees crown'd,   45
Whose loaded branches hide the lofty mound.
Such various wayes the spacious Alleys lead,
My doubtfull Muse knows not what path to tread:
Yonder the harvest of cold months laid up,
Gives a fresh coolness to the Royal Cup,   50
There Ice like Crystal, firm and never lost,
Tempers hot *July* with *Decembers* frost,
Winters dark prison, whence he cannot flie,
Though the warm Spring, his enemy draws nigh:
Strange! that extremes should thus preserve the snow,   55
High on the Alps, or in deep Caves below.
    Here a well-polisht Mall gives us the joy
To see our Prince his matchless force imploy;
His manly posture and his gracefull mine
Vigor and youth in all his motion seen,   60
His shape so lovely, and his limbs so strong
Confirm our hopes we shall obey him long:
No sooner has he toucht the flying ball,
But 'tis already more than half the mall;
And such a fury from his aim has got   65
As from a smoking Culverin 'twere shot.

    Near this my muse, what most delights her, sees,
A living Gallery of aged Trees,
Bold sons of earth that thrust their arms so high
As if once more they would invade the sky;   70
In such green Palaces the first Kings reign'd,
Slept in their shades, and Angels entertain'd:
With such old Counsellors they did advise
And by frequenting sacred Groves grew wise;

Free from th' impediments of light and noise 75
Man thus retir'd his nobler thoughts imploys:
Here CHARLS contrives the ordering of his States,
Here he resolves his neighb'ring Princes Fates:
What Nation shall have Peace, where War be made
Determin'd is in this oraculous shade; 80
The world from *India* to the frozen North,
Concern'd in what this solitude brings forth.
His Fancy objects from his view receives,
The prospect thought and Contemplation gives:
That seat of Empire here salutes his eye, 85
To which three Kingdomes do themselves apply,
The structure by a Prelate rais'd, *White-Hall*,
Built with the fortune of *Romes* Capitol;
Both disproportion'd to the present State
Of their proud founders, were approv'd by Fate; 90
From hence he does that Antique Pile behold,
Where Royal heads receive the sacred gold;
It gives them Crowns, and does their ashes keep;
There made like gods, like mortals there they sleep
Making the circle of their reign compleat, 95
Those suns of Empire, where they rise they set:
When others fell, this standing did presage
The Crown should triumph over popular rage,
Hard by that House where all our ills were shapt
Th' Auspicious Temple stood, and yet escap'd. 100
So snow on *Ætna* does unmelted lie,
Whence rowling flames and scatter'd cinders flie;
The distant Countrey in the ruine shares,
What falls from heav'n the burning mountains spares.
Next that capacious Hall, he sees, the room, 105
Where the whole Nation does for Justice come:
Under whose large roof flourishes the Gown,
And Judges grave on high Tribunals frown.

Here like the peoples Pastor he do's go,
His flock subjected to his view below;   110
On which reflecting in his mighty mind,
No private passion does indulgence find;
The pleasures of his youth suspended are,
And made a sacrifice to publick care;
Here free from Court compliances He walks,   115
And with himself, his best adviser, talks;
How peacefull Olive may his Temples shade,
For mending Laws, and for restoring trade;
Or how his Browes may be with Laurel charg'd
For Nations conquer'd and our bounds inlarg'd: 120
Of ancient Prudence here He ruminates,
Of rising Kingdoms and of falling States:
What Ruling Arts gave great *Augustus* fame,
And how *Alcides* purchas'd such a name:
His eyes upon his native Palace bent   125
Close by, suggest a greater argument,
His thoughts rise higher when he does reflect
On what the world may from that Star expect
Which at his birth appear'd to let us see
Day for his sake could with the Night agree;   130
A Prince on whom such different lights did smile,
Born the divided world to reconcile:
What ever Heaven or high extracted blood
Could promise or foretell, he will make good;
Reform these Nations, and improve them more,   135
Than this fair Park from what it was before.

*Note: in some versions, these lines occur between lines 66
and 67:*

        May that ill fate my Enemies befall   5
        To stand before his anger, or his ball.

iii. Richard Lovelace, "The Grasse-hopper. To my Noble Friend, Mr. Charles Cotton. Ode"

I.

Oh thou that swing'st upon the waving haire
    Of some well-filled Oaten Beard,
Drunke ev'ry night with a Delicious teare
    Dropt thee from Heav'n, where now th' art reard.

II.

The Joyes of Earth and Ayre are thine intire,   5
    That with thy feet and wings dost hop and flye;
And when thy Poppy workes thou dost retire
    To thy Carv'd Acron-bed to lye.

III.

Up with the Day, the Sun thou welcomst then,
    Sportst in the guilt-plats of his Beames,   10
And all these merry dayes mak'st merry men,
    Thy selfe, and Melancholy streames.

IV.

But ah the Sickle! Golden Eares are Cropt;
    Ceres and *Bacchus* bid good night;
Sharpe frosty fingers all your Flowr's have topt,   15
    And what sithes spar'd, Winds shave off quite.

V.

Poore verdant foole! and now green Ice! thy Joys
    Large and as lasting, as thy Peirch of Grasse,
Bid us lay in 'gainst Winter, Raine, and poize
    Their flouds, with an o'reflowing glasse.   20

VI.

Thou best of *Men* and *Friends*! we will create
    A Genuine Summer in each others breast;

And spite of this cold Time and frosen Fate
  Thaw us a warme seate to our rest.

VII.

Our sacred harthes shall burne eternally   25
  As Vestall Flames, the North-wind, he
Shall strike his frost-stretch'd Winges, dissolve and flye
  This *Aetna* in Epitome.

VIII.

Dropping *December* shall come weeping in,
  Bewayle th' usurping of his Raigne;   30
But when in show'rs of old Greeke we beginne,
  Shall crie, he hath his Crowne againe!

IX.

Night as cleare *Hesper* shall our Tapers whip
  From the light Casements where we play,
And the darke Hagge from her black mantle strip,   35
  And sticke there everlasting Day.

X.

Thus richer then untempted Kings are we,
  That asking nothing, nothing need:
Though Lord of all what seas imbrace; yet he
  That wants himselfe, is poore indeed.   40

MAJOR EDITIONS USED
AND CONSULTED

The texts used for the most frequently cited authors are "old-spelling" editions. For convenience sake, I have sometimes interpolated or substituted material in square brackets, and I have taken liberty with titles, especially in matters of italic usage.

Asterisks distinguish the editions used for quotations, determination of canon, etc.

Authors follow in alphabetical order of their surnames.

The text used for Waller has been most kindly provided by Philip R. Wikelund of Indiana University, who is preparing the definitive edition.

ANTHOLOGIES AND COLLECTIONS

(Three excellent anthologies are given first. The other titles offer collections, of various merits and kinds, possessing historical importance.)

Cavalier Poetry. Ed. Robin Skelton. London, 1970.
Jonson and the Cavaliers. Ed. Maurice Hussey. London, 1964.
The Tribe of Ben. Ed. A. C. Patridge. London, 1966; Columbia, S. C., 1970.

Cavalier and Puritan: Ballads and Broadsides Illustrating the Period of the Great Rebellion 1640-1660. Ed. Hyder E. Rollins. New York, 1923.
The Cavalier Songs and Ballads of England from 1642 to 1684. Ed. Charles Mackay. London, 1863.
Merry Songs and Ballads Prior to the Year A.D. 1800. Ed. John S. Farmer. 5 vols. N. p., 1897.
The Pepys Ballads. Ed. Hyder Edward Rollins. 8 vols. Cambridge, Mass., 1929-1932.
Seventeenth Century Songs now first printed from a Bodleian

Manuscript. Ed. John P. Cutts and Frank Kermode. Reading, 1956.

Songs and Lyrics from the English Playbooks. Ed. Frederick S. Boas. London, 1945.

Tudor and Stuart Love Songs. Ed. J. Potter Briscoe. London, 1902.

CHARLES COTTON (1630-1687)

*Poems on Several Occasions. London, 1689. (Including numerous translations and *The Battail of Yvry.*)

*The Genuine Works of Charles Cotton. London, 1715. (Including the burlesques, *The Wonders of the Peake,* and *The Planters Manual.*)

Poems of Charles Cotton, 1630-1687. Ed. John Beresford. London, 1923.

*Poems of Charles Cotton. Ed. John Buxton. London, 1958. (The first edition to make use of the manuscript in the Derby County Library.)

ABRAHAM COWLEY (1618-1667)

*The Works of Mr. Abraham Cowley. 9th ed. London, 1700.

Abraham Cowley. Poems. Ed. A. R. Waller. Cambridge, 1905.

Abraham Cowley. Essays, Plays, and Sundry Verse. Ed. A. R. Waller. Cambridge, 1906.

SIR JOHN DENHAM (1615-1669)

*The Poetical Works of Sir John Denham. Ed. Theodore Howard Banks. New Haven, 1928. (Superceded for *Cooper's Hill* by the following.)

*Expans'd Hieroglyphicks. A Critical Edition of . . . Cooper's Hill. Ed. Brendan O Hehir. Berkeley and Los Angeles, 1969.

WILLIAM HABINGTON (1605-1654)

*The Poems of William Habington. Ed. Kenneth Allott. London, 1948. (The excellent introduction and annotation in this edition will benefit all students of seventeenth-century poetry.)

ROBERT HERRICK (1591-1674)

*The Poetical Works of Robert Herrick. Ed. L. C. Martin. Oxford, 1956, and reprinted.
*The Complete Poetry of Robert Herrick. Ed. J. Max Patrick. New York, 1963, and reprinted.
(The University of Texas Library has a manuscript volume with copies of two poems by Herrick corrected, possibly by him.)

JAMES HOWELL (?1594-1666)

*Δενδρολογία. Dodona's Grove, or, the Vocall Forest. London, 1640.
*Δενδρολογία. Dodona's Grove, or the Vocall Forest. Second Part. London, 1650.
*Epistolae Ho-Elianae: Familiar Letters. 11th ed. London, 1754.

BEN JONSON (?1573-1637)

*Ben Jonson. Ed. C. H. Herford, Percy and Evelyn Simpson. 11 vols. Oxford, 1925-1952.
Poems of Ben Jonson. Ed. George Burke Johnston. London, 1954, and reprinted.
*The Complete Poetry of Ben Jonson. Ed. William B. Hunter, Jr. New York, 1963, and reprinted.
(The Yale edition will eventually replace the Herford-Simpson edition for the plays, and a new edition of the poems is being prepared by Ian Donaldson. Older editions, especially the three-volume Gifford-Cunningham, are still useful and can be found at moderate prices.)

MAJOR EDITIONS USED

RICHARD LOVELACE (1618-1658)

*The Poems of Richard Lovelace. Ed. C. H. Wilkinson. Oxford, 1930, and reprinted. (A two-volume edition [Oxford, 1925], also by Colonel Wilkinson, is more sumptuous and more ample but less easy to use.)

ANDREW MARVELL (1621-1678)

*The Poems and Letters of Andrew Marvell. 2 vols. Ed. H. M. Margoliouth. Oxford, 1927, and reprinted with some additions. (Pierre Legouis's revision of the commentary was in press as this book was completed.)

*The Poems of Andrew Marvell. Ed. Hugh Macdonald. London, 1952. (Superior text but not as complete as the preceding.)

THOMAS RANDOLPH (1605-1635)

*The Poems of Thomas Randolph. Ed. G. Thorn-Drury. London, 1929.

THOMAS STANLEY (1625-1678)

*The Poems and Translations of Thomas Stanley. Ed. Galbraith Miller Crump. Oxford, 1962.

SIR JOHN SUCKLING (1609-1642)

The Works of Sir John Suckling. Ed. A. Hamilton Thompson. London, 1910.

*The Works of Sir John Suckling. 2 vols. Ed. Thomas R. Clayton and L. A. Beaurline. Oxford, 1971.

HENRY VAUGHAN (1622-1695)

*The Works of Henry Vaughan. Ed. L. C. Martin. 2nd ed. Oxford, 1957, and reprinted.

*The Complete Poetry of Henry Vaughan. Ed. French Fogle. New York, 1964, and reprinted.

MAJOR EDITIONS USED

EDMUND WALLER (1606-1687)

*See the headnote to this section of the book.

The Works of Edmund Waller, Esq; in Verse and Prose. Ed. Elijah Fenton. London, 1744.

The Poems of Edmund Waller. Ed. G. Thorn-Drury. London, 1893.

INDEX

*The entries include names, topics, and titles (under their authors'
names) of works by Cavalier poets when the works are quoted. The
word "discussed" designates more extended discussion of a poem. Royalty are specified by their regnal names, nobility by their most familiar
titles, and classical persons by a single name except in cases where fuller
specification is necessary to avoid confusion.*

Agrippa, Marcus Vipsanius, 257
Allen, D. C., 5
Anacreon, 61, 88, 107, 108-11,
  148, 164, 286-88, 291
Antony, Mark, 99
Ariosto, Lodovico, 240
Aristotle, 68, 87, 250-51, 256-57,
  259, 263, 265, 279, 290-91,
  298, 309
Arnold, Matthew, 233
Ascham, Roger, 87
Ashcraft, Richard, 92
Aubin, Robert Arnold, 19
Aubigny, Katherine, Lady, 307
Aubrey, John, 184
Augustine, St., of Hippo, 94, 219
Augustus (Octavian) Caesar,
  28-29, 99, 119, 120, 137, 257

Bacon, Sir Francis, 87, 187, 240;
  on love, 208-209
Barclay, John, 68
Barker, Arthur, 175
Beaumont, Francis, 163, 260
Beaumont, Sir John, 75, 275
Beresford, John, 283
Bell, Robert, 21
Benlowes, Edward, 86
Blagden, Cyprian, 96
Blake, William, 136, 215
blazon, 228-29
Blitzer, Charles, 55
Bowman, Robert, 284
Brome, Alexander, 103

Bruno, Giordano, 216-17
Bunyan, John, 188, 191, 198,
  252, 300
Burlase, Sir William, 60
Burton, Robert, 108, 133, 165-66,
  205, 224-25, 230, 252-53,
  262; on love, 211-12
Butler, Samuel, 161, 187
Buxton, John, 283

Calvinism, 161, 189, 224
Camden, William, 260-65
Campion, Thomas, 103, 229
Carew, Thomas, 16, 28, 31, 38,
  78-84, 230, 243, 245, 255, 276;
  "To A. L. Perswasions to
  Love," 79, 106, discussed
  127-28; *Coelum Britannicum*,
  54-55; "In answer of an
  Elegiacall Letter," 83-84;
  "Mediocritie in love rejected,"
  85; A *Rapture*, discussed 80-82,
  251-52; "Song" ("Aske me no
  more"), 85-86, discussed
  135-37; "Song. Perswasions to
  enjoy," 106; "To Ben. Johnson,"
  269; "To Saxham," 198-99;
  "Upon a Mole in Celia's
  bosome," 80
Cartwright, Julia, 21
Cary, Sir Lucius: *see* Falkland
Cato the Elder, 94, 114
Catullus, 87, 103, 104, 108, 264
ceremony, and the good life, 49;

Editor of the matchless
Orinda," 302; "To the Pious
Memory of [Charles
Walbeoffe]," 180-81
Vaughan, Thomas, 219
Veen, Otto van, 121
verse epistle and friendship,
261-66
vir beatus, 76-84; *see also* good
life
vir bonus, 52-76; *see also* good life
Virgil, 61, 87, 108, 119, 151,
247, 257
vita beata, 76-84, 154, 158, 280;
see also good life
vita bona, 52-76; basis of,
52-54; *see also* good life

Waller, Edmund, 5, 10, 15-42,
56, 86, 107, 111, 137, 139,
168, 179, 188, 225, 227, 228,
238, 240, 244, 245, 249;
"A la Malade," 234-35; "At
Pens-Hurst," discussed 17-23;
"Chloris and Hilas," 235; "Of
a fair Lady playing with a
Snake," discussed 115-17; "Of

the last Verses," 16; *On St.*
James's Park, discussed 24-37;
Panegyrick to My Lord
Protector, A, 55; "Song" ("Go
lovely Rose"), discussed
39-41, 129; "To my Lord of
Falkland," 285; "To the King,
upon his . . . Return," 303
Walpole, Sir Robert, 77
Walton, Izaak, 5, 13, 44-45, 48,
49, 51, 52, 77, 79, 84, 102,
197, 298-99, 304-305, 306
Warder, Joseph, 289
Warwick, Sir Philip, 63, 65, 74
Wasserman, Earl R., 33, 167
Watson, George, 25
Webster, John, 56, 170, 200
Whalley, Peter, 12
Wicks (*or* Wickes, Weeks),
John, 131
Wild, Robert, 186
William III, 56
Wotton, Sir Henry, 304
Wycherley, William, 188

Zeno, 251

This book has been composed and printed by
Princeton University Press
Designed by Jan Lilly
Edited by Eve Hanle
Typography: Electra and Bodoni
Paper: Warren's Olde Style
Binding by The Maple Press Company